P9-CJM-938

WOMEN IN ATHENIAN LAW AND LIFE

WOMEN IN ATHENIAN LAW AND LIFE

ROGER JUST

ROUTLEDGE
London and New York

First published 1989
by Routledge
11 New Fetter Lane, London EC4P 4EE
29 West 35th Street, New York, NY 10001

Typeset in 10/12 Baskerville by Input Typesetting
Printed and bound in Great Britain by
Biddles Ltd, Guildford and King's Lynn

British Library Cataloguing in Publication Data
Just, Roger
Women in Athenian law and life
1. Ancient Greece Society. Role of women
I. Title
305.4′2′0938

ISBN 0 415 00346 6

Library of Congress Cataloging in Publication Data applied for

TO
GODFREY LIENHARDT

CONTENTS

PREFACE AND ACKNOWLEDGEMENTS

This book is essentially a work of synthesis. Moreover, since I write as a social anthropologist rather than as an ancient historian or a classicist, I am more than usually indebted to the scholarship of others. The informed reader, then, will find little here that is new, either evidence, or commentary on and interpretation of evidence. In writing this book I have had, however, two aims, which I hope will justify its appearance. Although a great deal has now been published about women in the ancient world (an invaluable bibliography of which up to 1981 was produced by Sarah Pomeroy), Athenian women have either been treated within the context of general works on women's position in antiquity (for example Pomeroy's own *Goddesses, Whores, Wives and Slaves*), or else they have been the subject, exclusively or in part, of more narrowly defined studies (for example David Schaps' *Economic Rights of Women in Ancient Greece*, or William Blake Tyrrell's excellent *Amazons, A Study in Athenian Mythmaking*). My first aim, then, has been to fill an obvious gap and to assemble within the covers of a single book a broad range of evidence concerning women in classical Athens presented in sufficient detail to afford the reader a comprehensive account of women's place in what was, after all, the most famous (and certainly the best documented) of the Greek city-states.

On the other hand, this book makes no claims to being encyclopedic. I have not included every mention of women in the Athenian sources. Nor have I attempted to review all that has subsequently been written about them. Indeed, with regard to secondary sources I have been highly selective, citing only those that I believe to be the most authoritative, or with which I most agree. For my second aim has been to present, step by step and somewhat in the manner of an ethnography, an argument and a sustained interpretation in which each aspect of women's position and representation is related to the other, so that what emerges is an account that is not only comprehensive, but also coherent. If, then, this work has any claim to originality, it lies in the hope that its whole is greater than the sum of its parts.

I have tried not to overburden with scholarly apparatus a book intended to be accessible to the general reader. I trust, however, that the many scholars on whose work it is constructed will find themselves duly acknowledged throughout the text. I am nevertheless aware that in a book written intermittently over a long period, and which deals with information most of which could now be considered 'public domain', it is possible that some reference or form of words which ought to have been acknowledged has passed unremarked. If this is at all the case, let me offer an unreserved apology.

Three works without which this book could never have been written require special mention. First, in matters relating to the interpretation of the law, I have seldom strayed from A. R. W. Harrison's monumental *The Law of Athens*. For the most part my contribution has been to embed Harrison's comments and interpretations within a more general sociological context. Second, any reader familiar with the writings of Sir Kenneth Dover will realize the extent of my debt to his works, particularly *Greek Popular Morality in the Time of Plato and Aristotle*, not only for the sources it gathers together, but also for the interpretations it supplies. Finally, in terms of its overall direction, this book owes much to John Gould's seminal article, 'Law, custom and myth: aspects of the social position of women in classical Athens', of which, in many ways, it is an expansion.

This book started life in 1976 as a B. Litt. thesis in anthropology at the Institute of Social Anthropology, Oxford. I am deeply grateful to G.E.M. de Ste Croix, then Tutor in Ancient History at New College, who was for some time my supervisor, and who was willing (as always) to give time and advice to someone who, from the ancient historian's point of view, might well have seemed a rank outsider. Within anthropology, my supervisor was Godfrey Lienhardt, an Africanist of repute, but whose judgements on things Greek were as shrewd as on everything else. The thesis was examined by J. K. Davies, then of Oriel College, Oxford, and John Campbell, of St Antony's College, Oxford. I am grateful for the encouragement that both of them offered me at the time, and it was subsequently under Dr Campbell's guidance that I embarked on field work in rural Greece and the study of modern Greek society which has sustained me ever since. Sir Kenneth Dover was kind enough to read some early revisions of the thesis, and I am grateful for his generous comments and advice. Sally Humphreys took an interest in my work from an early stage and has continued to prompt and encourage me over the years. I remain grateful for her advice and her example.

At Melbourne University, George Gellie most generously read through a draft of the work, and there are few people whose judgements I would more respect. In the final stages, and in the best spirit of co-operation, my colleague Robin Jackson read the entire typescript and saved me from more infelicities than I care to recall. Finally, I owe special thanks to Angela Khoury, who not only typed this book onto the computer, but had to keep on typing my continual changes to it.

It remains to say that although this work is built on those of others, faults in the final construction are certainly my own.

Roger Just
Melbourne, 1988

1

INTRODUCTION: THE 'PROBLEM' OF WOMEN

I

The study of women in classical Athens presents us with a dilemma. As A.W. Gomme remarked many years ago, 'There is, in fact, no literature, no art of any country, in which women are more prominent, more carefully studied and with more interest, than in the tragedy, sculpture, and painting of fifth-century Athens' (1925: 4). Moreover, there exists a considerable body of evidence, mainly in the form of law-court speeches, which allows us to reconstruct (though not without the inevitable gaps and uncertainties) the position of women within the political, legal, social and economic structures of the Athenian *polis*. It would seem, then, that social historians had a wealth of material at their disposal. Yet the fact remains that for all practical purposes there is nothing which represents the authentic voice of women themselves. Euripides' Medea may speak passionately for women, but it is still Euripides who does the talking. Contrary to the recorded comments of Athenian men, the women of Athens have kept a prudent silence. Further, compared with our knowledge of public institutions and political and military history, the private and day-to-day lives of the Athenians remain relatively opaque, and it is in this context that we might have hoped best to register the role of women, for they make scant appearance in the chronicles of Athens' greatness. Either, then, we know what men said about women and how they represented them, or we know very little about them at all; and when we look to determine the position of women in Athens we can claim to be determining only what Athenian men thought about women, and how rules and regulations constructed by men sought to define and locate women within a male conception of society (Gould

1980:38–9). This is a distressing situation, particularly at a time when a general attempt is being made to reassess the contribution of women to society and history; yet it is a situation whose several consequences must be accepted.

II

Problems of evidence (and of its omissions) are the special concern of historians, and nowhere are historians more exercised by them than in the study of ancient history. But given the particular difficulties attendant on studying women, it is worth noting certain developments within the anthropology of contemporary societies; for despite radically different conditions of investigation, there too strangely similar problems obtain.

It has been suggested with some justification that the minor role attributed to women in most ethnographies can be seen as the result of certain prejudices, or at least presuppositions, on the part of the ethnographer about the essentially masculine nature of society (and this whether the ethnographer happened to be male or female). However, in an important article originally published in 1972 (and from which I borrow the title for this chapter) Edwin Ardener argued that this minor role had little or nothing to do with the actual amount of information recorded about women, or even with the amount of interest shown in the position and activities of women. Rather, it had to do with the perspective within which that information was cast; for any prejudices or presuppositions the ethnographer might have had about the essentially masculine nature of society, and consequently about the peripheral role of women, were likely to be shared and reinforced by the very observations of the society under study. It seems that within society it is a male view which predominates, and that it is thus a male view of society which is most readily forthcoming. The ethnographer in the field, with all his or her 'information' in full view, still tends to see the society through male eyes because it is the men of the society who are the most willing and capable of providing an articulate account of their lives; an account in which women have, precisely, 'a place'. Indeed, as Ardener says, when attention turns to women, both ethnographer and 'his people' are likely to be engaged in a similar process of 'bird-watching' whereby women become passive characters in a masculine social drama rather than being seen as at the centre of their own interpreted world (Ardener 1975: 1–4).[1]

Nevertheless, what anthropologists now accuse themselves of is not so much a failure to have recorded the social truth about women, as a failure to have seen beyond one social truth about women located in a reality constructed by men. What current criticisms of the traditional anthropological treatment of women have highlighted is not so much the mistakenness of presenting a male view of women, as the mistakenness of not recognizing its relativity as a male view. And it follows that the rectification of this situation does not lie in discarding the male view of society in favour of a search for some elusive objectivity, but in supplementing that view with the female alternative. The male view of society still retains its significance, for social reality is a social construct and what people think themselves and others to be remains a primary object of social enquiry.

It may well have been, then, that beyond the dominant ideology of the male, which purports to account for society in its totality, there existed in Athens another social reality constructed by women in which not only their own role and nature, but also those of men, might have been construed in a significantly different fashion. Ideally, this would constitute a large part of the subject-matter in any investigation of Athenian women; but the evidence for it is not recoverable, and though I would certainly not wish to argue that classicists and historians are any the less prone to making sexist assumptions than the anthropologists of contemporary societies, it is scarcely their fault that the evidence from antiquity tends in certain areas to neglect women (in itself a salient fact) and in other areas to represent them in a manner which no doubt gives them very much less than was their due. On the other hand, what historians continually do confront, and what they can respect as a quite legitimate object of study, is precisely the prejudices, or should one say cultural assumptions, of the male society which has chosen to record its views for posterity in the enduring authority of written texts. Although all of what is known about women comes from the representations and ordinances of men, and although this evidence will not allow the discovery of the whole truth about women – about how they felt, about how they saw Athenian society and their place within it, or even very much about what they did within the confines of their private and domestic lives – it will nevertheless allow the discovery of something of the way in which Athenian men thought of women, of the place allocated to them in the male construction of society, and of the attributes and characteristics with which they were thereby credited. In discovering this, like most ethnographers (though with very much less choice), the historian can still claim to be recording an important, but 'partial' in every sense of the word, truth about women: not what they were, but what men saw them to be. This is to limit the study of Athenian women and to reorient its findings; but it is by no means to invalidate that study.

III

The particular nature of the evidence, its status as a register of male ideas and values rather than as a direct and intimate record of the female condition, has more effect, however, than merely to impose certain limits on the study of Athenian women. The difference between a naive view of society as a set of objective phenomena to be determined, examined, interpreted, and judged, and society as a construct already replete with judgements and meanings given it by those who live out its reality, lies at the heart of any attempt to discuss Athenian women, and explains the confusions created by many early attempts.

When Gomme published his famous essay in 1925 it was to counteract a nineteenth-century orthodoxy that in classical Athens women lived lives of cloistered confinement verging on 'oriental seclusion', that they were legally, politically, economically, and socially subjugated and suppressed, and that they were treated with an indifference approaching contempt.[2] Gomme's challenge to this view has been upheld by a number of later scholars, notably Kitto (1951), Seltman (1956), and later Richter (1971). As a corrective to so bleak a view of women's position it had great value; but to argue whether Athenian women were held in contempt or were honoured and cherished members of the community was to set most subsequent discussion of the position of women in Athens within the terms of a debate that could not advance.[3] To put it bluntly, the question which most classicists asked themselves, even if their answer remained cautiously inconclusive, was, 'Did the Athenians treat their women well or not?' or 'Was woman's position in Athens a happy one or not?' Admittedly this is an obvious and even important question to ask; but it is not a question it is possible to answer (Gould 1980: 39).

As Sarah Pomeroy pointed out in her bibliography on women in antiquity (1973: 141–3)[4] and again in her book *Goddesses, Whores, Wives and Slaves* (1975: 58–60), divergence of opinion on this question related not only to a subjective interpretation of the available evidence, but also to a subjective selection of the genres of evidence taken into account. The 'pessimists' view was based largely on a reading of the Athenian orators, on philosophers' and moralists' writings, and on what little can be pieced together from various prose sources and from comedy about Athenian daily life. The 'optimists' view sprang from a consideration of Athenian art and drama and from a recognition of the prominent place of women within them. Adherence to one or other of the opposed views then necessitated the mutual charge that the evidence on which the opposition's claims were based was either unrepresentative or unimportant. Thus while Gomme and his followers would argue that the legally defined position of women in Attic law had little or nothing to do with the respect and honour granted them, W.K. Lacey, for example, writes in his book on the family in classical Greece: 'Amongst the intentional omissions of this book are large-scale references to Greek Tragedy. . . . What the characters say [in tragedy] . . . has no independent value for telling us about society, though very often it will support what we know from other sources to be true' (1968: 10). But although the selection of evidence was an important factor in the controversy and raises certain very real questions to which I shall return, it is not the decisive factor. The dispute over the status of women resulted from a more fundamental error.

For the most part, questions about the position of women in Athens were posed from the outset in moral or evaluative terms. Were women despised and held in contempt, or were they honoured and cherished? But it is not possible to read through the corpus of Athenian writings and substantiate in any empirical fashion from direct statements about the matter whether Athenian men as a whole liked or disliked women, whether they went around honouring or despising them. Nor is this surprising. There is a body of explicitly misogynic Greek (though not necessarily Athenian or classical) literature. Hesiod and Semonides of Amorgos come immediately to mind. So does Hippolytos' lengthy monologue in

Euripides' tragedy. But Hippolytos' speech must be understood within its dramatic context and, as Gomme correctly pointed out (1925: 8–9), for every explicitly misogynic statement extracted from literature one can find another to the effect that there is no greater blessing for a man than a good woman – a sentiment to which even Hesiod subscribed. This being the case, the recourse was naturally to evidence other than direct and unambiguous expressions of affection or hostility. From such indirect evidence classicists then set themselves the task of deducing whether the Athenians honoured or despised their women. But these deductions perforce entailed a series of *a priori*, personal, and culturally relative judgements about what sort of behaviour towards women, what sort of conditions for women, and, indeed, what sort of characterizations of women would be indicative of an attitude of admiration or disparagement. The trading of opinion was considerable. The same applies even to the question of 'oppression'. From the available evidence it can certainly be said that in Athens a woman's life was in most areas a much more restricted one than a man's; that she was not allowed to do, or simply did not do, many of the things that a man did. And, by the way, I am by no means arguing that this could not have constituted oppression. Certainly by our standards it would. But before one could confidently talk of oppression, or of treatment constitutive of contempt, it would be necessary to know whether the restrictions imposed upon women contravened or frustrated their own desires; whether women resented their situation and felt themselves underrated or even despised. It would be necessary to know something of their own assessment of their situation – and this is not known. What took its place were classicists' opinions about how one ought to treat women – and they, of course, were various.[5]

Being treated badly or well, being honoured or despised, must depend on the culturally and historically specific estimations of those who are or were themselves the inhabitants of the society or period under consideration.[6] Such estimations cannot be supplied by proxy, for they do not concern matters which permit of objective determination on universal grounds. It seems clear that Athenian men, about whose opinions we know something, did not think themselves contemptuous of women. This being so, the acceptability or otherwise of their treatment of women would become an interesting and possible debate only if something were known of what

Athenian women themselves thought of their situation. But given the state of the evidence the most that can be done is occasionally to remind ourselves that not everything to which women might now, and with great justification, object was necessarily resented by the women of Athens, and conversely to appreciate that historical changes have altered, and will continue to alter, people's consciousness of themselves, of others, and of gender roles within society. But perhaps the most unfortunate effect of the long and essentially futile debate over women's 'despised' or 'honoured' state in Athens was that it served to obscure what was a worthwhile and possible field of investigation: namely, the very nature of those 'prejudices' or cultural assumptions which lie embedded in a variety of contexts from law through to literature and which together reveal what the meaning of 'woman' or femininity was for the Athenians.

IV

In a somewhat rhetorical passage Seltman wrote at the outset of his book *Women in Antiquity*: 'one must be precise about terms of reference. "Woman." There is no need to attempt a definition. We are always with them, and they with us. Fortunately' (1956: 15). Yet is the definition self-evident? Can we presume to know in any universal sense what is meant by 'woman' (or for that matter what is meant by 'man')? True, both the Greek term '*gyne*' and the English term 'woman' refer to a human being of female gender. But the connotations of both terms are far more extensive, and there is no certainty that what was brought to mind by the word '*gyne*' in fifth- and fourth-century BC Athens was the same as what is brought to mind by the word 'woman' in contemporary western society – not, that is, unless the meaning of the words is restricted to nothing other than a human being who is biologically female. Rather than taking the definition of woman as universally given, and then proceeding to see whether she was 'rightly' or 'wrongly' characterized and treated within an alien culture, it is, I believe, more profitable to reverse the process and to attempt to see how the characterizations and treatment of women in an alien culture went towards defining what a woman actually was. In short, what must be investigated is the semantic field of 'woman' within the confines of a particular society and culture. That, increasingly, has been the general direction of the more recent studies of women in ancient society – a direction which runs parallel to much contemporary anthropological writing, now that women's studies have lost their initial stridency and begun to investigate the rich complexity of gender differentiation throughout history and society.[7] It is also the direction of this book.

Needless to say I shall not be attempting to find or to formulate any neat definition of the Athenian idea of woman. Rather, the task will be to trace through the evidence of those various systems of thought and behaviour, those institutions and representations within which women were granted a place to see how and whether they might allow the gradual construction of some coherent picture of what the Athenians thought woman to be. This cannot be done quickly. First, I intend to examine woman's incorporation (or lack of incorporation) into the Athenian *polis*, and her role within kinship organization and family structure. Here I shall be concerned not only with the legal rules which defined her capabilities and incapabilities, but also with her economic status and the degree of her participation in the less formally defined areas of social life. Second, I wish to consider the comments made in fifth- and fourth-century writings about the 'nature' or 'character' of women. Here I will also be concerned with those characteristics, both behavioural and 'psychological', which were thought ideally to be the prerogative of men – for it will be necessary to see in what way women were thought to be different from men in order to appreciate the meaning of those characteristics which men attributed to women. Finally, I shall review some of the symbolic associations and connections into which women entered in the rituals of the Athenian state and also in the stories, the myths, of Greece on whose traditions the Athenian playwrights drew. It is from such a range of sources and different contexts that I hope a coherent picture of the Athenians' conception of women may emerge.

I shall not, however, be attempting to evaluate the justice of this conception, its 'correctness' with respect to our own moral persuasions. On the other hand, what I shall continually be dealing with are moral evaluations and notions of morality, both explicit and implicit – those of a past and alien society. According to Thoukydides' account, Perikles' funeral oration, delivered at the end of the first year of the Peloponnesian War, contained the following words:

> If I am also to speak about feminine virtue, referring to those of you who will from now on be widows, I can say all I have to say in one short word of advice. Your great glory is not to be inferior to the way nature made you; and the greatest glory is hers who is least talked about by men, whether in praise or in blame.
>
> (Thoukydides: 2.45)

Richter (1971) has argued that the sentiments of the speech should be attributed to Thoukydides the historian rather than to Perikles the statesman since Thoukydides is 'prejudiced' against women and regularly ignores them in his history; that the sentiments could scarcely carry a genuinely misogynic connotation if they were Periklean, since clearly Perikles was no woman-hater (Richter reminds us of Perikles' notorious relationship with the courtesan Aspasia, and of the anecdote that on his death-bed Perikles sheepishly admitted having kept through the years an amulet some woman had given him); and that the advice to women given in this passage is but another expression of that characteristic Hellenic ideal of *sophrosyne* ('discretion', 'prudence') since in an oration of this sort any reminder that public familiarity with a respectable woman's private life might compromise her would be apt. I am not sure why a 'prejudice' on Perikles' part would be sociologically important, but can be dismissed if expressed by Thoukydides. (Indeed, that Thoukydides hardly mentions women in his history seems quite in accord with their general exclusion from the public domain.)[8] But, whatever the case, the substance of Richter's comments functions only within the framework of an illusory argument – an argument that assumes that if Perikles said what he said and meant what he said then he is open to the charge of despising women. That Perikles was no woman-hater is clear; but whether Perikles loved or hated women is not, except for his biographer, an important question. The real question must be *what*, for the Athenian male, were these women whom Perikles, or Thoukydides, or any number of other Athenian men were at liberty *either* to cherish *or* to despise according to their individual persuasions; what, for Perikles, Thoukydides and all the rest, constituted a 'good woman' whom ideally they ought to honour and respect, or a 'bad woman' whom they ought to deride. The answer given here is clear, though incomplete, and must be respected: 'The greatest glory is hers who is least talked about by men, whether in praise or in blame.'

V

In trying to reconstitute the Athenian notion of women from a range of evidence I shall, however, be forced to deal with one particular problem raised by the established debate over women's status – a problem to which Pomeroy has drawn attention: the apparent discrepancy between woman's role in day-to-day life and her role in art and tragedy. As Gould (1980: 40) has argued, it can be shown that contrary to the assertions of Gomme and his followers the social conventions which the heroines of tragedy obey (or are criticized for disobeying) do not differ radically from those to which the women of fifth- and fourth-century Athens appear to have been subject. Nevertheless, those who have adduced the evidence of tragedy and art have still raised a substantial problem: how is one to reconcile the very forcefulness of those awesome women who appear in tragedy, and the magnitude of their actions continually celebrated in drama, with the values of a society which otherwise would appear to have consigned women to the background? Athenian art and the public performances of dramatic festivals were just as much a part of the social reality of Athens as were the haggles over inheritances which we read of in the orators or someone's wife sitting at home in the women's quarters spinning wool. Even if the impact which the women of tragedy have on the lives of men does not accord with the passive role they appear to have had in daily life, we cannot dismiss these dramatic representations as irrelevant to an understanding of Athenian society. There must have been some connection between the women of tragedy and the women of Athens. Art, and the society which produces it, are not separate phenomena; the former is an integral part of the latter, an element of the same historical reality.

It is here that I think an approach which concentrates on what the (male) idea or notion of woman was may prove productive, for both sides in the above debate employ the same basic assumption, viz., that the evidence from tragedy is, or could be, valid only inasmuch as it faithfully describes the social conditions of Athens. Thus Lacey, for example, will allow the evidence of tragedy only when it will support what we know from other sources to be true, while

Gomme, for his part, forthrightly states that the highest comedy and the highest tragedy are both 'but representations of life' (1925: 5). It seems to me that this is by no means so, or rather, that it is naive to suppose that the only valid relationship artistic expression can have with society is realistically to recreate the social conditions which created it. Athenian drama does not have to *describe* Athenian society or Athenian women to be an important register of the ideas, beliefs, values, and attitudes of that society and of their implications for women. I would suggest that the question of the degree to which the women of tragedy accurately represented the women of Athens should be side-stepped, and that instead what should be examined is the degree to which certain ideas or notions about the nature of women, explicit or implicit in other evidence, might also be expressed, perhaps in rather different guise, in the representations of tragedy; that the focus should be shifted from the purely empirical to the ideological, and from points of factual contiguity to points of conceptual similarity. As I remarked at the outset, all we know about women in Athens comes from the comments of men, or is derived from rules and regulations whereby men sought to define and locate the position of women within their conception of society. By concentrating on ideas rather than 'facts' it is at least possible to place the evidence from tragedy alongside the evidence from other sources in a manner which, though it may fail to reveal what Athenian women were truly like, may yet allow the discovery of that 'partial' truth of which I spoke.

VI

One last comment is in order. It should be made clear that in attempting to build up a picture of Athenian women I shall be producing a synthesis of common-place ideas and attitudes and I shall be working at a high level of generality. Individuals have always disagreed with the collective representations of their society and, even at the collective level, the Athenians' response to women appears at times ambivalent and contradictory. Nevertheless, it is still possible to arrive at some picture of what it was to which Athenian men were responding even if it represents only the sort of general image which, modified, differently evaluated, or even rejected by particular individuals in their particular circumstances, had its place in their minds as the collective, cultural, and stereotypical idea of women.[9] This picture has its own interest. If, for example, the women of Athens are continually characterized as drunkards in the comedies of Aristophanes, we do not have to believe that Athenian women were all drunkards, or even that most or a significant minority of Athenian women were drunkards. Nor do we have to believe that most Athenian men thought that their own mothers, or wives, or sisters, or daughters, or even the mothers, wives, sisters and daughters of their friends, were all drunkards. We can respect the probability that we are dealing with comic exaggeration or comic calumny. Nevertheless, the fact that drunkenness was a stock charge against women in comedy needs consideration. After all, different *categories* of people are mocked in comedy for different reasons. Whatever the truth of the matter, and whatever the individual Athenian's opinion of his own womenfolk, why was it that when the Athenians wanted to laugh about *women* they called them drunkards? Some general notion of the female character is being appealed to. It is that general notion I wish to recover.

2

POLITICS

I

Women in Athens possessed no active political rights. They could neither speak nor vote in the *ekklesia*, the citizen assembly, nor could they attend its meetings. Further they were unable to hold any administrative or executive position within the secular organization of the state (including that of juror in the popular courts).[1] In the Greek sense of the word, they were not citizens – in Aristotle's definition, participants in the offices and honours of the state.[2]

These general disabilities are so well known that they scarcely require elaboration; nor is there anything peculiarly Athenian about the exclusion of women from politics. No Greek state ever enfranchised women, and for the most part the general participation of women in politics has had to wait until the twentieth century. Yet if the mere fact of women's exclusion from politics is not in itself particularly noteworthy (since it is but another example of what, historically, has been almost universally the case), the importance of that exclusion within the context of Athenian society must still be stressed, for 'politics' in Athens denoted something much more extensive than, and rather different from, what is understood by the term today.

II

Athens was a democracy, and it was a democracy of a radical kind. From the reforms of Ephialtes in 462 BC down to the oligarchic revolutions after 322 BC, with only a few years' interruption, it was the *demos*, the people, who held power directly as the collective administrators and legislators of the state. The practical reality of the situation was perhaps sometimes different. Wealth, birth, and influence continued to count; aristocratic and propertied families could still dominate; and, by virtue of his personal standing, a single man, Perikles, could determine for nearly fifteen years the policies of the entire state. Yet these qualifications do not alter the fact that Athens' *form* of government was radically democratic. Every Athenian citizen (unless he had been deprived of his rights) was eligible to speak and to vote in the assembly, which was the sovereign body of the state. Every Athenian citizen over the age of thirty was eligible to hold public office, and, with the exception of military commands and later of some high financial offices, the complex administration of the whole state was run by amateur boards of magistrates chosen annually by lot. Through the popular courts, the administration of justice was also firmly in the hands of the people (Hopper 1957; Finley 1975: 70–82; Jones 1957: 99–113). The government was the people. But a question remains: who were 'the people'?

In fact the limits of the Athenian democracy can readily be seen without having to take into account the *de facto* appropriation of political power by individuals or by groups within the citizen body. Its limits are clearly evident in terms of the sections of Athens' permanent population which lay *de iure* outside the citizen body, for the idea of unrestricted suffrage was not part of Athenian (or Greek) democratic theory. As Aristotle wrote in his *Politics*:

> We do not for a moment accept the idea that we must call citizens all persons whose presence is necessary for the existence of the state.
>
> (1278a 1)

Those whose presence was necessary for the existence of the state, of the *polis*, but who were excluded from its governance, who were not 'citizens', *politai*, included, of course, all women and children. But there were other groups, too, who were politically deprived: resident aliens or 'metics', who had a defined legal status in Athens and who were officially registered there as inhabitants; freedmen, whose status was largely assimilated with that of the metics and who, for practical purposes, can be included in the same category; and, finally, slaves (Hopper 1957).

Unfortunately population figures are among those basic facts of

Athenian history about which no more than educated guesses can be made; but when Athenians spoke of citizens, they meant *free adult males of recognized Athenian parentage*, and during the democratic period that group could have accounted for only a relatively small proportion of Athens' total population. Indeed, even of the adult *male* population, citizens were perhaps scarcely in the majority, and citizenship remained an hereditary and a jealously guarded privilege.[3]

The scrutiny of those claiming citizenship was close. The penalties for fraudulently claiming citizenship were severe. The emotions aroused by the questioning of citizen status were strong. A speech composed by Demosthenes well illustrates the situation. An examination of the deme lists (the only registers of the citizen body) had been ordered by decree of the assembly in 346 BC. A certain Euxitheos was then expelled from his deme by vote of its members and thereby deprived of his citizenship. The expulsion had resulted from a charge brought by a certain Euboulides that Euxitheos was not of Athenian parentage. In the speech Euxitheos appeals against the decision of his fellow demesmen by taking his case before a court in Athens. I quote from the beginning and the end of the speech:

> In my opinion it is your duty to treat with severity those who are proved to be aliens (*xenoi*), who, without either having won your consent or asked for it have by stealth and violence come to participate in your religious rites (*hiera*) and your common privileges, but to bring help and deliverance to those who have met with misfortune and who can prove that they are citizens; for you should consider how pitiful above all others would be the plight of us whose rights have been denied, if, when we might properly sit with you as those exacting the penalty, we should be numbered among those who pay it, and should unjustly be condemned along with them because of the passion which this subject arouses . . .
>
> (Demosthenes 57 [*Euboulides*]: 3; Loeb trans., modified)

> I for my part am left without a father, but for my mother's sake I beg and beseech you to settle this trial so as to restore to me the right to bury her in our ancestral tomb. . . . Do not deny me this; do not make me a man without a country; do not cut me off from such a host of relatives and bring me to utter

ruin. Rather than abandon them, I will kill myself that at least I may be buried by them in my country.

(ibid., 70)

It should be noted that even in his delicate situation Euxitheos affirms the 'them' and 'us' dichotomy of non-citizen and citizen. Indeed, it is precisely to this dichotomy that Euxitheos appeals. He would be one of the judges, not one of the judged. Further, the strength of emotion aroused by the question of citizenship is explicitly referred to. But Euxitheos' passionate determination to defend his citizen status is shown best by his final words. It would be easy to see them simply as a rhetorical plea designed to sway the hearts of the jurors – and to an extent they are surely that. But it should not be forgotten that in challenging the decision of the demesmen and in bringing his case before a court in Athens, Euxitheos was running the risk, in the event of an adverse decision, of having his property confiscated and of being sold into slavery. In this context the reference to suicide should not lightly be dismissed.[4]

But why was citizenship of such importance? Why did individuals run such risks in vehemently maintaining it (or in illegally usurping it)? Why did the Athenians so severely guard it? The advantages of being free as opposed to being a slave scarcely need comment; but had Euxitheos accepted the decision of his demesmen he would still have been a free man, though reduced to the status of a metic. What, then, for a free man was the importance of citizenship? What distinguished a citizen's life in Athens?

The answers to these questions are not entirely self-evident, if only the material or utilitarian advantages of citizenship are taken into account. There is no doubt that certain tangible benefits accrued to citizens which were not shared by the free non-citizen population. The most important of these was the right to own real estate for, generally speaking, only citizens could possess land and houses in Attica; and land in Athens, as in the ancient world in general, was the paradigm form of wealth.[5] Euxitheos came from a rural deme. Loss of his citizenship, and consequently loss of his holdings, might well have meant his economic ruin. Further, the inability of the wealthy non-citizen to convert his capital into land was a social as well as an economic barrier, for land meant prestige as well as security. But this scarcely explains the importance placed on citizenship. There were many wealthy non-citizens, and very many poor citizens. In the arts, crafts, and trades the non-citizens prospered. True, they paid special taxes; they were not eligible for the occasional free distributions of corn made to citizens and they could not receive the small sums paid for attending the assembly, the council, or the courts, or for holding public office. But, on the whole, the non-citizens' economic disadvantages were slight; and if we turn to their legal rights then, for most practical purposes, they were on an equal footing with citizens. Their person and property were protected by law; they could even prosecute a citizen in the courts; and in the city's daily life metics and citizens appear to have mingled on easy and familiar terms (Ehrenberg 1974a: 147–64; Harrison 1968: 187–99).

Despite this situation, however, and despite grants of citizenship occasionally made by special vote of the assembly, the exclusive nature of the citizen body was actually strengthened during the fifth and fourth centuries. A law proposed by Perikles in 451/0 and reinstituted in the archonship of Euklides in 403/2 ordained that a citizen had to be of Athenian parentage on both sides, his mother's as well as his father's. Further, in the course of the fourth century, it became illegal for a non-Athenian to marry an Athenian, and the penalties for transgression were severe: slavery and the confiscation of property. In short, we appear to be confronted with a situation in which the free non-citizen population of Athens was neither persecuted nor discriminated against, in which metics lived in practical equality with citizens, in which, with the major exception of being able to own land, they were free to prosper economically, but in which the distinction between citizen and non-citizen was rigidly maintained and the exclusiveness of the Athenian citizen body reaffirmed.

To understand why citizenship was so jealously guarded, then, I think it is necessary to look beyond its purely practical advantages and to look in two directions which may at first seem somewhat divergent.

I shall quote two passages at some length. The first is from Thoukydides' account of Perikles' funeral oration, perhaps the most famous declaration of Athenian ideals:

> Our constitution is called a democracy because power is in the hands of the people, not of a minority. When it is a question of settling private disputes, everybody is equal before the law:

> when it is a question of putting one person before another in positions of public responsibility, what counts is not membership of a particular class, but the actual ability which the man possesses. No one, so long as he has it in him to be of service to the state, is kept in political obscurity because of poverty. . . . Here, each individual is interested not only in his own affairs, but in the affairs of state as well. Even those who are most occupied with their business are very well informed on general politics. This is a peculiarity of ours: we do not say that a man who takes no interest in politics is a man who minds his own business; we say that he has no business here at all.
>
> (Thoukydides: 2.37 and 40; Rex Warner trans., modified)

The second is from a speech in the Demosthenic corpus in which a certain Theomnestos and his brother-in-law, Apollodoros, are prosecuting the Athenian citizen Stephanos and the Corinthian courtesan Neaira for living together as man and wife, contrary to the laws of Athens. Theomnestos begins the address:

> Stephanos here sought to deprive me of my relatives contrary to your laws and your decrees; so I too have come before you to prove that Stephanos is living with an alien woman contrary to the laws; that he introduced children not his own to the members of his phratry and deme; that he has given in marriage the daughters of prostitutes as though they were his own; that he is guilty of impiety towards the gods; and that he nullifies the right of your people to bestow its own favours if it chooses to admit anyone to citizenship. For who will any longer seek to win this reward from you and to undergo heavy expense and much trouble in order to become a citizen when he can get what he wants from Stephanos at less expense, assuming that the result for him is to be the same?
>
> (Demosthenes 59 [*Neaira*]: 13; Loeb trans., modified)

The two passages are neatly complementary in stressing different, but by no means opposed, aspects of the meaning of citizenship and politics in Athens. Perikles' speech is a eulogy of the democracy, of a form of government which allowed everyone, regardless of background or wealth, to participate actively and to the best of his abilities in the running of the state; indeed, of a form of government which made it everyone's duty to do so. No matter how much the practical reality of the matter differed, the right to participation and the call to participation were there. But when Perikles talks of everyone, he means every citizen, every adult male of recognized Athenian parentage. Let me then state the obvious. What the democracy reserved for its citizens, and what above all else distinguished them, was precisely the exercise of political power – political power which was embodied in more than a vote; which was embodied in the right to daily participation in government. And we should note that such participation was not limited to those living in the urban centre with easy access to the assembly or the courts. Other arenas existed. When Euxitheos, from whose speech I have already quoted, appealed to a court in Athens against the charge laid against him by Euboulides which had resulted in his loss of citizenship, what was being presented to that court was part of a history of rivalry and public competition for honour which had been played out at the deme level among a group of only about eighty members – but none the less savagely for that (Humphreys 1983: 23–4). Citizenship was prized in Athens because to be a citizen, however humble, was to have a share in the public life and the rule of the *polis* and to have access at a number of levels to the public competition for honour and prestige. It was to be a privileged person – and it might be supposed that such privileges become jealous possessions.

In part Theomnestos' speech shows similar concerns. He prosecutes Stephanos out of long-standing enmity. For the citizen, private quarrels are played out in a public context. Further, his brother-in-law Apollodoros, who delivers the bulk of the speech to follow, was himself a new-made citizen – hence both his desire to enter into public debate and the stress placed on the debasement of citizenship by Stephanos' actions. Yet if it was the comparatively recent innovation of the radical democracy which made being a citizen of such value, there were, I think, other more conservative forces which kept the citizen body exclusive and to which this speech also makes appeal.

What Theomnestos (and later Apollodoros) describe is a society which viewed itself in terms of a natural unity in which the solidarity of its members was expressed through the idioms of kinship and shared religion. Apollodoros could join the ranks of the citizens through his wealth and public benefactions; but the Athenians did not see society as merely the aggregate of so many autonomous individuals bound together only by their rational acceptance of a common good. One was a man because one was a member of a *polis* – a *polis* which was as much a social and religious unity as in our sense of the word a 'political' one. A man's social existence derived from his situation within a network of what remained essentially kinship connections; connections that supplied him with his social and religious identity and life; connections that extended to embrace both the dead and the yet-to-be-born as well as the living; connections which, ideally, were inherited by birth not contracted by choice. Citizenship, the right to hold and exercise political power, was more than a compact whereby the common acceptance of certain rights and duties by individuals was creative of a community; it was the consequence of a man's membership of a community already well defined by bonds of kinship and religion.[6]

In this context Stephanos' introduction into his phratry and deme of boys whom he had not fathered, and his giving in marriage to Athenian citizens of daughters who were not his own, was not only a breach of the law – it was an offence against those very principles in terms of which the Athenian community was ideally defined; principles which were certainly capable of manipulation and modification as historical and political circumstances demanded, but which nevertheless remained fundamental to the Athenians' conception of the *polis*. It is worth recalling the scrutiny which every Athenian magistrate, including every member of the council, had to undergo before taking office, in which he was required to establish by way of witnesses the name and deme of his father, of his father's father, of his mother and of his mother's mother; whether he maintained the cult of Ancestral Apollo and Household Zeus; whether he had family tombs and where they were; and whether he treated his parents well.[7]

Indeed, it is worth recalling Euxitheos' words:

> do not make me a man without a country; do not cut me off from such a host of relatives and bring me to utter ruin. Rather than abandon them, I will kill myself that at least I may be buried by them in my own country.

III

In considering the importance of politics and citizenship in Athens, I may appear already to have strayed some distance from the specific subject of women. But both factors to which I have drawn attention – the radical nature of the democracy in which all citizens, but only citizens, had the right to direct participation in government, and the continued expression of the Athenian community's cohesion in terms of the idioms of kinship and religion – have, in differing fashion, substantial implications for the study of women. It is between these two aspects of Athenian political life that much of the nature of women's social role is revealed.

There are enough indications to be certain that some women at least were aware of public issues, and that in Athens men might, in the privacy of the home, discuss with their wives and daughters what had taken place in the courts or in the assembly (Lacey 1968: 173–4). It is also conceivable that some women influenced their husbands' political decisions, for the domineering wife, or mother, was as much a joke in Athens as in our own society. When, in Xenophon's *Memorabilia*, young Lamprokles complains to Sokrates of his mother's vile temper, Sokrates, wishing to uphold the gratitude and respect owed to all parents, enquires,

> Which do you think is the harder to bear: a wild beast's brutality, or a mother's?
>
> I should say a mother's if she's like mine.
>
> (Xenophon, *Memorabilia*: 2.2.7 Loeb trans., modified)

And in a specifically political context one thinks immediately of Perikles' mistress, Aspasia, as she is portrayed in Plutarch's *Life of Perikles*. But Aspasia's notoriety and the popular resentment her supposed influence aroused should also be remembered – a resentment transmuted into mockery by comedy.[8] Politics, the domain of public life which so occupied Athens' men, was not women's proper place. In fact women were scarcely given nominal membership of the citizen body. The feminine form of 'citizen', i.e. *politis*, does occur, but in general when it was necessary to distinguish the mother, or wife, or daughter of an Athenian citizen from other women (as it often was), then the somewhat odd term *aste* (city-woman) was used.[9] Nor should it be forgotten how much of the humour of Aristophanes' *Lysistrata, Ekklesiazusai* and, to an extent,

Thesmophoriazusai are derived from masculine fantasy about how women would act if ever they had access to the city's public life (Dover 1974: 96); or the shock created in the assembly when a certain Pamphilos announced that public funds were being embezzled by a man and *woman*, only then to reveal that he was in fact referring to the effeminacy of a man.[10] And when, in Xenophon's *Oikonomikos*, Ischomachos claims to Sokrates' surprise that:

> 'I have often been singled out before now, Sokrates, and condemned to suffer punishment or to pay damages.'
> 'By whom?'
> 'By my wife.'
> (Xenophon, *Oikonomikos*: 11.25; Loeb trans., modified)

What is supposed to make the passage humorously appealing is that Ischomachos has created within his own household a microcosm of the Athenian democracy in which his wife is capable of impeaching him in just the same way as an Athenian citizen could in the 'real world' of politics outside. But it is only a game, for in that real world of politics women had no place, as Perikles' celebrated words make clear.

Although it is unlikely that the position of women in Athens underwent any substantial decline as a result of the democracy, the observation sometimes made that women's position was worse during the democracy than in earlier periods, and worse in Athens than in other states, may have a truth.[11] In narrowly oligarchic, aristocratic, or monarchic states, women who belonged to the elite have often wielded considerable power, even if illegitimately; on the other hand, since the bulk of the population, whether male or female, possessed no political rights, 'politics' was not something which in general distinguished men from women. But in the Athenian democracy there were no thrones from behind which women could rule (unless one cares to credit the complaints against Aspasia), while the access that every adult Athenian male had to the offices and honours of the state sharply distinguished the citizen's life from that of his wife or daughter (cf. Pomeroy 1975: 78).

Whatever women's power and influence within the private domestic sphere, Athens' democratic system allowed men to enter into a public life of the city from which women continued to be excluded, and from which vantage point the domestic confines within which women remained, though perhaps little different from before, must have seemed truly domestic and confined. The democracy accentuated a major disparity between the lives of men and women; a disparity which can be seen in terms of the continual contrast between public and private which runs throughout Athenian thought.[12] Further, and a closely related point, Athenian women found themselves on one side of what had become an important boundary even for Athens' male population: the boundary between those who were entitled by birth to participate in the running of the state, and those who were not. It would be untrue to say that women were thus assimilated with metics or with slaves. Athenian women were certainly to be distinguished from those non-Athenian women living in Athens, and the difference between being free and being slave was quite as radical for women as it was for men. Nevertheless, the democracy meant that Athenian wives and daughters, by contrast with their men, were placed as a group alongside those others who, like metics, were outsiders, or who, like slaves, were always officially subject to rule rather than being capable of becoming rulers themselves. This, as we shall see, affected the male idea of women in a number of contexts, for, in Vidal-Naquet's words, the Athenian *polis* was both a 'citizens' club' and a 'men's club' (1981a: 188). By definition women fell outside both.

IV

But let me now view matters from the other direction. As a 'citizens' club', whose members all shared in the governance of the state, the Athenian *polis* excluded women. But as a closed community bound together by ties of kinship and religion it most certainly included women, both in their own right and, within a largely bilateral inheritance system, as channels through which economic, religious, and most importantly political rights were transferred and transmitted between men.

The participation of mothers, wives, sisters, and daughters of Athenian citizens was vital to the religious life of the city. They took part in the rites and cults of individual households, *oikoi*, of the various divisions of the state – the demes, phratries, tribes, and *gene* – and of the *polis* itself. It was, for example, a woman from the *genos* of the Eteoboutadai who held the position of priestess of Athena Polias (Athena of the city); and while it was the citizen selected to act as Archon Basileus (King Archon) for the year who presided over all the ancestral festivals, his wife, given the title of Basilissa, presided over state ceremonies with him, and also represented the wife of the god in the annual ritual which celebrated the mystical marriage of Dionysos.[13] It should be noted that on the religious plane the wives of Athenian citizens formed quite as exclusive a group as their husbands. They alone could participate in the great women's festival of the *Thesmophoria*. As Marcel Detienne points out in *The Gardens of Adonis* (1977: 78), at the end of the fifth century BC a woman's participation in the festival of the *Thesmophoria* was a legally valid proof that she was legitimately married to an Athenian citizen in full possession of his political rights. Female slaves, the wives of metics and foreigners, and concubines, and courtesans were all excluded from the festival.

More generally, the obligations and respect owed to kin, and especially to parents, and which were sanctioned by law as well as enshrined in morality, applied to women as much as men. A law ascribed to Solon and quoted in Aischines 1 (*Timarchos*) deprived of his civic rights any man 'who beats his father or his mother, or fails to provide for them or support them'. These same provisions are entailed in the scrutiny of magistrates to which I have already referred. Again, Euxitheos' appeal, quoted above, was to be permitted to bury his *mother* in the family's ancestral tomb and, as I have suggested, this was probably intended as an emotionally compelling appeal to the male jurors. Indeed, sentimental appeals based on the welfare of mothers, wives, sisters, and daughters were a commonplace tactic of law-court oratory (Lacey 1968: 175–6).

But beyond being the objects of such moral concern, women were fundamental to the organization and structure of the *polis*, even if their role there was largely passive. Rights of religious and economic inheritance were transmitted through women as much as through men, and it was through connections via women as well as through connections via men that an individual was situated within the kinship structures from which his social identity derived. Although women were not themselves 'citizens', the maintenance of the Athenian *polis* as a closed community of citizens depended on rules and regulations which recognized women and which incorporated women in the official structures of the family and the state. Indeed, as will be seen, the very definition of an Athenian citizen involved not only his being born of an Athenian father, but also of an Athenian mother properly given in marriage by her kin. As soon as the emphasis is shifted from the possession of active political rights to a consideration of the manner in which the body politic was itself constituted, then women appear as much a part of the state as men.

In what follows, then, I shall attempt to give a more detailed account of the social and legal regulations involving or affecting women. In many cases I shall be giving only a brief and partial review of matters which, largely owing to the nature of the evidence, are complex, and for all further points of legal detail the reader is advised to turn to A.R.W. Harrison, *The Law of Athens: The Family and Property* (1968). My consistent aim, however, will be to draw attention to the two aspects of women's position outlined above: on the one hand, women's exclusion from any active participation in the secular affairs of the state; on the other hand, women's inclusion within the structure of the *polis* as members necessary for its perpetuation and cohesion as a closed community. Inasmuch as the democracy was made up of citizens all of whom had access to the offices and honours of the state, the complete exclusion of women from this group deprived them of what had become of enormous importance in the life of the male Athenian, and so entered into the definition of feminity in a new or at least much more pronounced way. Perhaps this is to say no more than that in a society in which public life is an important matter, women's exclusion from it should be seen as a correspondingly more important element in their definition – but it was an element which, as I hope to show, affected the very conception of what a 'woman' was. On the other hand, inasmuch as citizenship and political rights remained an hereditary privilege within a community whose unity was still expressed in terms of the idioms of kinship and religion, the maintenance of Athens' exclusive '*club de citoyens*' and the bonds which linked one citizen to another depended largely on the social role which women were made to play. In this sense they were without doubt integral members of the *polis*.

3

LEGAL CAPABILITIES

I

A woman's life long supervision by a guardian, her *kyrios*, summarizes her status in Athenian law. She was not considered a legally competent, autonomous individual responsible for her own actions or capable of determining her own interests. As Harrison observes:

> There can be no doubt that a woman remained under some sort of tutelage during the whole of her life. She could not enter into any but the most trifling contract, she could not engage her own hand in marriage, and she could not plead her own case in court. In all these relations action was taken on her behalf by her *kyrios*, and this was so during her whole life.
>
> (1968: 108)

As an adjective the word *'kyrios'* means literally 'having power or authority over', 'capable', 'authorized', 'official'; as a substantive it means 'master', 'controller', 'possessor'. A woman's *kyrios* was responsible for her domicile, her maintenance, her upbringing as a child, and her general welfare. He also represented her and acted on her behalf in all undertakings subject to legal provisions or from which legal consequences derived – notably her marriage.

Briefly, a girl's *kyrios* was in the first instance her father. If her father was dead, her homopatric brother or paternal grandfather acted as her *kyrios*. When she was married her husband became her kyrios. If she was widowed or divorced she returned to the charge of her original *kyrios*, if possible to be engaged in marriage by him to somebody else. If she was pregnant by her deceased husband she could remain (and perhaps had to remain) under the guardianship of her husband's heir until the child was born. Finally, if she was widowed with sons who were minors, she probably could elect to remain within the household of her late husband under the guardianship of her sons' appointed guardian(s) until her sons came of age; alternatively, if her sons were already of age, she could place herself under their guardianship (Harrison 1968: 109–11).

I shall return in more detail to the appointment of a woman's *kyrios* and his rights over her person and property in later chapters; for the moment it is sufficient to point out that *kyrieia* (guardianship) over a woman was exercised by one or other of her male relatives, or her husband, or possibly their temporary appointee, throughout her entire life. Two closely related aspects of this situation should be stressed, for not only did the law thereby make a woman incapable of asserting her own rights and acting as an adult individual in matters of any legal consequence, it also presumed that a woman, at least an Athenian woman, was always to be situated within a family or *oikos* under the guardianship of its head. Here it should be noted that the role of a kyrios entailed more than the supervision of women. The term *kyrios*, denoted the head of a household, an *oikos*, which comprised women, male minors, and property (including slaves). The *oikos* was, in short, the basic Athenian family unit (Lacey 1968: 15ff.). Any change in a woman's status as a result of marriage, divorce or widowhood also resulted in a change in the person of her *kyrios*; but in each case the change also meant her incorporation into a new household or *oikos*, or her return to her natal *oikos*. Thus not only her social role, but her very legal identity, tied her to the family. She was always the member of an *oikos*, and always under the protection of whatever male was head of that *oikos*.

The derivative nature of a woman's social and legal identity which stemmed from her familial relationships with men is clearly reflected, or rather manifested, by the manner in which public reference to her was usually made. Such passages as the following are typical of law-court language:

> The real sons of Euktemon the father of Philoktemon, namely Philoktemon himself, Ergamenes and Hegemon, and his two daughters and their mother, the wife of Euktemon and the daughter of Meixiades of Kephisia, are well known to all their relatives and to the members of their phratry and to most of

> the members of their deme, and they shall testify to you; but no one is aware or ever heard a word during Euktemon's life time of his having married any other wife who became mother by him of our opponents.
>
> (Isaios 6 [*Philoktemon*]: 10–11; Loeb trans., modified)

> he (Apollodoros) made a will in case anything happened to him and devised his property to Archedamos' daughter, his own sister and my mother, providing for her marriage with Lakratides who has now become hierophant.
>
> (Isaios 7 [*Apollodoros*]: 9; Loeb trans., modified)

Men are specified by their proper names; women are specified by their relationships with men (Schaps 1977).

It is worth noting that there was no taboo on publicly referring to a woman by name. The names of respectable mothers, wives, sisters, and daughters of Athenian citizens do occur; nevertheless, the normal practice was to refer to a woman as so-and-so's mother, wife, sister, or daughter, and we know the names of remarkably few of the many women mentioned in law-court proceedings. Nor was this because women were unimportant in such contexts. In many cases their position was crucial. In the second of the above-quoted passages, for example, the daughter of Archedamos, the mother of the speaker and the sister of the late Apollodoros, was named at one stage by Apollodoros as the heir to his property; yet her name is not repeated in court. The simplest explanation for this is that the specification of a woman in terms of her familial relationships with men was quite sufficient to establish her identity. For the importance of her identity, indeed its sole legal importance, lay precisely in her relationship with men and her situation within a household. The only categories of women regularly referred to by their proper names were *hetairai* and *pallakai*, prostitutes and concubines, who for the most part were slaves, freed women, or foreigners, and who by definition lay outside the boundaries of the Athenian kinship system and thus could not reasonably be referred to in any other way (Schaps 1977).[1]

From the Athenian point of view a woman's *kyrios* was her protector rather than her master: certainly morally, and to an extent legally, he was required to attend to her welfare (Harrison 1968: 111–12; Lacey 1968: 107). It would thus be possible to see a woman's *kyrios* simply as the male relative who at any particular time was required to provide her with domestic support and representation in a society in which women had little role in the extra-domestic world – were it not for the fact that despite her legal incapabilities a woman could inherit considerable wealth and be at least the nominal claimant of an estate. This did involve her in worldly affairs, but as the passive victim of male ambitions.

As the equivalent of legal minors, women could not administer or control any of the property which might be inherited in their name, or even the money with which they had been dowered (Harrison 1968: 236, 52). But Athenian law was at further pains to limit the economic transactions and capabilities of women. A law referred to in Isaios 10 [*Aristarchos*]: 10 reads: 'For a minor is not allowed to make a will. The law expressly forbids either a minor to engage in any contract, or a woman beyond the value of one *medimnos* of barley'. Kuenen-Janssens (1941) has attempted to determine what the value of one *medimnos* of barley was. She concludes that it was not negligible; perhaps enough to feed a family for five or six days, and certainly enough to account for the petty trading activities conducted by women which we hear of in Aristophanes' comedies and elsewhere. Its worth cannot, however, be specified with any precision, and it is doubtful that it was intended as an exact limitation. Despite her arithmetic, Kuenen-Janssens expresses the conviction that, 'one *medimnos* of barley indicated a rather vague idea, informal as Greek law often was' (1941: 212). But the law was no dead letter. It is humorously reversed to apply to men when, in Aristophanes' *Ekklesiazusai*, the women take over the state.[2] And even if the law did not make women dependent for their every penny on the consent of men, women could not engage in any major transactions involving the substance of their dowries or their inherited estates. This being so, a woman's *kyrios* might find himself in control of a considerable fortune vested in the person of his charge, but which he administered on her behalf and, so far as it can be judged, at his discretion. In litigation, a woman's *kyrios* might also find himself pressing the rights of his charge to an estate and property which he would enjoy or, conversely, attacking the rights of another woman to such property, or even the rights of guardianship that another *kyrios* had over a woman.[3] In all such cases, women's complete incompetence before the law made them dependent on the goodwill and capacities of the *kyrioi* who represented them.

It can be presumed that in any criminal proceedings in which a woman was the accused, or in which, having been offended by someone outside the family, she was the plaintiff, family honour and solidarity if not personal loyalty would come into play, and a *kyrios* would do his best on behalf of his charge.[4] In property claims a woman's *kyrios* would in most cases have every reason to uphold her rights, since to neglect them would be tantamount to neglecting his own. But if the speaker of Isaios 10 [*Aristarchos*] is to be believed, and the conjectures of Wyse (1904: 276) about the situation behind Isaios 3 [*Pyrrhos*] are correct, then it seems quite possible that a girl's valid claims to an inheritance could be thwarted by a family conspiracy to which her own *kyrios* was party, and about which consequently she could do very little.[5] This raises the question of just how vulnerable a woman's lack of any independent legal status made her, and though the evidence will not allow a general answer, it is interesting to note the nature of the provisions which Athenian law made to bypass the role of the *kyrios* in protecting those who were not legally competent to protect themselves. It is also of interest to note the rationale which lay behind such provisions.

II

Among his other duties the eponymous archon was charged with the general oversight of orphans, heiresses, and widows who claimed to be pregnant by their late husbands. In his capacity as a magistrate, he had the right summarily to impose fines up to a fixed limit on anyone who committed an offence against such people. If a larger fine was in order the archon could cite the offenders before the court over which he presided (Harrison 1971: 4, 8). Thus if a *kyrios* failed to provide proper maintenance for his ward(s) the archon could exact it from him. Further, the process of *eisangelia*, 'impeachment', was available against anyone, including a guardian, who wronged an orphan or an heiress. Such wrongs, *kakoseis*, included general neglect (Harrison 1968: 117–21).

The employment of the *eisangelia* in this context is of some interest. For the most part legal charges in Athens fell into one of two classes, *dikai* and *graphai*, which, very roughly, might be considered as private suits and criminal complaints, though in fact the difference was procedural. A *dike* could be instigated only by the plaintiff or by his/her guardian; a *graphe*, on the other hand, could be brought by *ho boulomenos*, that is to say by whosoever wished (of those legally competent to do so) (Harrison 1971:74–8). Like the more normal *graphe*, an *eisangelia* could be brought by whosoever wished, but its principal distinguishing feature was that the prosecutor ran no risk. He had neither to deposit a court fee to be forfeit should his prosecution fail, nor was he subject to any penalty if he failed to secure at least a certain proportion of the jurors' votes. The other contexts in which the *eisangelia* was available involved the most serious of charges: those against official misconduct by arbitrators, and those against anyone deemed to be threatening the state by his acts – attempts to overthrow the constitution, the taking of bribes by orators, and that unfortunate political charge of 'deceiving the people' (Harrison 1971: 50–9).

The availability of the *eisangelia* for the protection of orphans and heiresses would seem to indicate the gravity with which any offence against them was considered. Certainly it placed such offences among those whose prosecution the state made every effort to facilitate. The penalty in the case of conviction was fixed by *timesis*, a process of assessment and counter-assessment between prosecutor and defendant, and if the defendant was *kyrios* of a maltreated orphan or heiress he was also deprived of his guardianship over them (Harrison 1968: 118). It would be tempting, then, to see such legal provisions as indicative of a benevolent aspect of Athenian law which sought to offer women, as legal minors, special protection beyond that afforded by their *kyrioi* and which, indeed, would protect them against any possible mistreatment by their *kyrioi*. The categories of women covered by these provisions, however, show the real concern of the law, as does the use of the process of *eisangelia* itself.

The two categories of women who, along with orphans, enjoyed the special protection of the law were widows claiming to be pregnant by their late husbands, and heiresses. A male orphan, that is to say a boy whose *father* was dead, would on coming of age succeed to his father's *oikos* and inherit both its wealth and its family cult, thus maintaining its continuity. A pregnant widow was the possible bearer of a male heir for her late husband's *oikos*, while heiresses, *epikleroi* (a special category of women which will be discussed in detail later), although incapable of providing the desired continuity of the *oikos* by their own person, were again potential suppliers of male heirs. For any son that an *epikleros* bore could become the direct heir to his maternal grandfather's household. Thus the only women who benefited by the archon's oversight were those through whom a male heir could be supplied for an *oikos* which was temporarily bereft. The concern of the law in this respect was to ensure the correct transmission of property and, importantly, of religious rights and duties via direct *male* descent, rather than to compensate women for their legal incapabilities. The extra personal protection that pregnant widows and *epikleroi* received against possible mistreatment by their *kyrioi* was incidental to that concern. Further, it is the strength of that concern rather than any particular solicitude over the treatment of women by their *kyrioi* or others that explains the availability of the *eisangelia.* In a society still expressing its unity through the idioms of kinship and family, interfering with the correct succession of the constitutive *oikoi* of the *polis* was tantamount to threatening the state itself.

There is, of course, no particular reason to think that the majority of Athenian women and girls were not adequately protected by their *kyrioi.* Respect and loyalty to the family were central to Athenian morality. Furthermore, a woman's *kyrios* was normally, though not necessarily a close male relative or her husband. Inasmuch as difficulties for a woman could arise, however, it is worth noting some further legal provisions from which women might have benefited – though their workings are by no means clear.

There were two specific charges which could be brought against a guardian: a *dike epitropes* ('trusteeship suit') and a *dike sitou* ('maintenance suit'). The problem here is that both were *'dikai',* i.e. 'private suits', which could be instigated only by the plaintiff himself or by his/her guardian. If a charge was to be brought against a guardian, it is thus difficult to see by whom it could be instigated. In the case of a male minor the probability is that the *'dike epitropes'* could be brought only by the plaintiff himself when he came of age – as in the famous suits of Demosthenes against his guardians.[6] If the ward was female, however, and thus in a permanent state of legal minority then, as Harrison admits, 'we are entirely in the dark as to the age at which and the person through whom a *dike epitropes* could be set on foot' (1968: 119). The same holds with the *dike sitou.* When he came of age Demosthenes charged one of his mother's guardians with having failed to support her adequately, despite being in control of her dowry. But how a widow, for example, could sue her guardian if she had no son or before her son was of age remains unclear. Such a situtation does, however, raise the question of multiple *kyrieia* over a woman, and consequently the protection a woman might have had against mistreatment by one *kyrios* by appealing to another *kyrios*. For it seems fairly certain that, although upon her marriage a woman passed into the *kyrieia* of her husband, the *kyrios* of her natal *oikos* still retained certain residual rights over her, notably the right to dissolve her marriage (Harrison 1968: 30–2). If this is so, then a woman could appeal, for example, to her father against any possible mistreatment of her person or property by her husband, and the situation might be resolved by divorce or the threat of divorce (which, personal considerations aside, would deprive the husband of the enjoyment of his wife's dowry). Finally, if we consider criminal actions, then as Harrison remarks, 'if a woman was, for example, killed and her husband as *kyrios* declined to take proceedings – or if indeed he had killed her himself – we may conjecture that her original *kyrios* could act, but this is nothing but conjecture' (1968: 32).

III

In fact the question of a woman's vulnerability due to her lack of independent legal status cannot readily be answered in general terms. Such cases of mistreatment or of fraud as occur in law-court material are by definition the extreme ones concerning matters, usually financial matters, weighty enough to have been taken before a court. And what was generally at issue was in any case the competing claims of *men* to the wealth of women in their charge or to wealth which they might obtain through their connections with women. The effect which women's legal incapacity had on their day-to-day lives is not readily apparent from such material, though some attempt to deal with this question will be made in later chapters. Nevertheless, the provisions of Athenian law discussed above exemplify two fundamental and complementary aspects of women's position within Athenian society: on the one hand, the impossibility of any major independent action being undertaken by her in the public sphere; on the other hand, the assumption of her permanent incorporation within and protection by the *oikos*, the household, under the governance of its male *kyrios* and head. But although it is certain that a woman could not instigate or conduct any legal proceedings on her own behalf, her very familiarity with the internal affairs of the household within which she permanently dwelt would seem to make her an ideal witness in any case which required evidence from such a quarter. Lacey has stated that, 'In legal actions the Athenians allowed their womenfolk to give evidence. . . . The family in fact was its own record office and in this field the women's evidence was apparently reckoned by an Athenian court to be as good as the men's' (1968: 174). Since women were excluded from direct participation in other areas of secular public life, it is worth examining in a little detail whether they did hold the position which Lacey suggests in that institution so characteristic of the Athenian democracy – the courts.

It is generally accepted that the actual giving of evidence to an Athenian court (orally in the fifth century, deposited in writing in the fourth century) was a privilege restricted to free adult males (not necessarily citizens), but that an exception was made in the case of homicide where women and slaves could give evidence against, but not on behalf of, the accused (Harrison 1971: 136–7). MacDowell, however, argued that the evidence will not allow us to decide whether the exception held or not (1963: 102ff.; cf. Harrison 1971: 136). The question of homicide cases aside, Bonner and Smith have stated that, 'Normally the evidence of a woman could be presented by her *kyrios*' (1938: 131; cf. Harrison 1971: 137 n.1). There is, I think, some difference between a woman giving her own evidence and the evidence of a woman being given by her *kyrios*, which suggests that her evidence was not quite 'reckoned by an Athenian court to be as good as the men's'. Certainly it had a different status, since women did not attend the courts. And while Bonner and Smith's statement is in principle correct, it is worth scrutinizing the sources to see what the evidence given by a woman's *kyrios* amounted to.

In the Athenian courts a man was his own advocate, and related his version of events. Witnesses did not relate what they knew to have happened but were merely called on by the speaker to corroborate his account. Much information was therefore given without direct confirmation and this might include what a woman was deemed to have said. Of this there are examples.[7] Further, the actual evidence of witnesses might also include what a woman was deemed to have said or done. Of this too there are examples.[8] However, that a woman's own evidence could be presented to the court by her *kyrios* rests on two passages: Demosthenes 57 [*Euboulides*]: 67, and Isaios 12 [*Euphiletos*]: 5.

In Demosthenes 57 the speaker, Euxitheos, cites the relatives, the *oikeioi*, who have given testimony about his father's identity and citizenship:

> first, four (male) cousins; then the son of a cousin; then those who are married to the female cousins.
>
> (Demosthenes 57 [*Euboulides*]: 67; Loeb trans., modified)

All that is said here, however, is that the husbands of Euxitheos' female cousins have given evidence, and since these husbands were themselves Euxitheos' *oikeioi* (for the term *oikeioi* included affines), to argue that the evidence they gave was necessarily their wives' evidence rather their own seems specious. Further, when Euxitheos cites the relatives who have given evidence of his mother's identity and Athenian birth, he lists the following:

> First, a nephew; then two sons of her other nephew; then the son of a cousin; then the sons of Protomachos who was my mother's former husband; then Eunikos of Cholargos who married my sister the daughter of Protomachos; then my sister's son.
>
> (ibid., 68)

It is difficult to see all this as anything other than the direct evidence of men.

The second passage from Isaios 12 is more suggestive. Again the case concerns citizenship, that of a certain Euphiletos, and the speech is delivered by Euphiletos' homopatric half-brother who, in the following, argues the reliability of the evidence produced:

> For observe in the first place that the husbands of our sisters would never have given false evidence in favour of Euphiletos; for his mother had become stepmother to our sisters, and it is usual for differences to exist between stepmothers and the daughters of a former marriage, so that, if their stepmother had born Euphiletos to any man other than our father, our sisters would never have allowed their husbands to give evidence in his favour.
>
> (Isaios 12 [*Euphiletos*]: 5; Loeb trans., modified)

Here it seems fairly clear that the evidence which had been

produced by the husbands of the sisters of Euphiletos was indeed that of their wives. But was it legally considered women's evidence? It was the husbands who still stood to be prosecuted for false witness if they were proved to have related what was untrue. Further, it is difficult to imagine that in general men could be counted on to act as mere channels through which women could introduce their evidence into a court. Again I think it is simply the case that the husbands of Euphiletos' sisters gave evidence in their own name of what their wives had told them; and to argue the stronger point that this passage proves that women had not only the right to give evidence via their *kyrioi* but also to stop evidence of which they did not approve being given in their name seems to me quite unsound. If women did want to deny such evidence what possible means did they have of forwarding their denial, since it was their *kyrioi* who spoke for them? All that can be argued from this passage, and all that is actually stated by the speaker, is that on purely informal and *de facto* grounds a wife would discourage by whatever means at her disposal a husband from making statements in court attributed to her of which she did not approve. The speaker is referring to the recognized situations of domestic and conjugal life, not to any legal provision – though, of course, the nature of that reference and its introduction into a court of law is in itself of considerable significance in revealing the Athenians' attitudes towards women. They were not recognized as legally competent persons outside the home; but they were certainly recognized as having some authority within it, and there was no embarrassment about publicly admitting as much.

The one major qualification which it seems necessary to make with respect to women's legal incapability relates to the giving of oaths. This, however, was a rather special procedure. Normally the evidence of an Athenian witness was unsworn (though he could still be liable for a charge of false witness). Nevertheless, either the litigant himself or a witness on his behalf could volunteer to take an oath as to the veracity of his statements in order to render them more persuasive (Harrison 1971: 150). Usually a challenge was issued to the opposing party to take a similar oath. Statements under oath could not, however, be made without the agreement of both parties; but even if no oaths were taken since one party declined to agree, the content of the oath accompanied by a statement of the party's willingness to swear to it could still be related to the court (Harrison 1971: 152).

According to Demosthenes 39 [*Boiotos I*] and Demosthenes 40

[*Boiotos II*] (in which the story is repeated), two sons of a certain woman, Plangon, were rejected by their putative father Mantias as not being his own and legitimate sons. The speaker is a third son of Mantias recognized by his father:

> Plangon . . . having in conjunction with Menekles [apparently a well-known rogue] laid a snare for my father and deceived him by an oath that amongst all mankind is held to be the greatest and most awful, agreed that if she was paid thirty *minai* she would get her brothers to adopt these men [the two rejected sons] and that, on her part, if my father were to challenge her before the arbitrator to swear that the children were in very truth his own sons, then she would decline the challenge.
>
> (Demosthenes 40 [*Boiotos II*]: 10; Loeb trans., modified)

Unfortunately for Mantias, this ruse was foiled and Plangon double-crossed him:

> When these terms had been accepted . . . he went to meet her before the arbitrator, and Plangon, contrary to all she had agreed to do, accepted the challenge and swore in the Delphinion an oath that was the very opposite of the former one. . . . Thus my father was forced on account of his own challenge to abide by the arbitrator's award, but he was indignant at what had been done and took the matter heavily to heart, and would not even agree to admit these men [his putative sons] into his house; but he was compelled to introduce them into his phratry [i.e. as his legitimate sons].
>
> (ibid.)

In another case (concerning Euphiletos' citizenship, already referred to above) we are told that Euphiletos' mother too was willing to swear an oath before arbitrators, although here it seems the oath was never actually sworn:

> And in addition to the depositions, men of the jury . . . the mother of Euphiletos, who is admitted by our opponents to be a citizen-woman (*aste*), expressed before the arbitrators her willingness to swear an oath at the Delphinion that Euphiletos here was the issue of herself and our father; and who had better means of knowing than she?
>
> (Isaios 12 [*Euphiletos*]: 9; Loeb trans., modified)

Further, in Demosthenes 55 the content of an oath which the speaker's mother was willing to swear apparently was read to the court:

> I tendered an oath to my mother and challenged them (the opponents) to have my mother swear in the same terms. Take, please, the depositions and the challenge.
>
> (Demosthenes 55 [*Kallikles*]: 27; Loeb trans., modified)

In these cases evidence which was without doubt the direct evidence of women was given due recognition, especially in the case of Mantias and his putative sons where Plangon's evidence was decisive. Mantias was compelled to admit the two young men to his phratry as his legitimate offspring. But several points should be noted. First, this evidence was in the form of an oath. The sanctions it invoked were religious as much as civic, and in the first two cases above we hear that the oaths were, or would have been, given at the Delphinion, the sanctuary of Apollo. In this context we are dealing perhaps not so much with an exceptional instance of women's participation in the secular affairs of men, as with the presentation to a court by men of a form of evidence expected to carry weight by virtue of its very sanctity. Lacey's failure to distinguish between sworn and unsworn evidence may be partially responsible for his view that women's evidence was as good as men's (Harrison 1971: 137 n.1). Secondly, although the efficacy of Plangon's oath in securing the acceptance of her sons should in no way be dismissed as unimportant, it was only the exceptional circumstance of her broken compact with Mantias that allowed an oath contrary to Mantias' interests ever being sworn. Mantias permitted the swearing of an oath because he thought he had predetermined its content; but in the other cases above the oaths were not taken, and had therefore only rhetorical value, because the opposing party declined to agreed to them. In court the opposing parties were always men, and this being the case the possibility of a woman's direct evidence (even under oath) being presented to the court still depended on the willingness of her *kyrios* to tender her an oath and on the willingness of the opposing party to allow the oath to be tendered.

In sum, all that can safely be said is that if a woman was in possession of pertinent knowledge to the advantage of her *kyrios* as either a litigant or as a witness in court, then that knowledge would find its way before the court and be given due recognition. Further, in allowing this, the Athenians certainly admitted women's interests in and familiarity with the internal affairs of the *oikos*, and men might well have been expected as a matter of course to relate what their womenfolk had told them. It would still, however, be most misleading to say that women had the right to have their views presented in court or that in this context their legal incapacity was any the less.

IV

In this section I have been concerned to examine the degree to which women participated in what might be defined as the world of men – of *politai* and of public affairs. The minimal nature of this participation should be obvious. Athenian political life excluded women from the secular offices and honours of the state; Athenian law legislated women's inability to act as independent agents in all matters subject to legal provisions from which legal consequences derived. The complement to this should, however, be equally obvious. A woman's place was within the home, within the family; and though she may not have been physically constrained to remain there, that was her realm of competence. Her role in secular affairs was purely domestic, an adjunct to the private life of the citizen within whose *oikos* she was situated, under whose control and protection she was placed, and through whom all her dealings with the outside public world were mediated. Yet though woman's own place may have been restricted to the domestic realm, in a society in which, despite a growing separation between private and public affairs (Humphreys 1983: 22–32), family and kinship were still basic to the definition of the body politic, woman's seclusion within the *oikos* did not mean that her status was without significance to the state. In the next two chapters, which are concerned with marriage, and with property and the family, I shall, therefore, reverse the perspective and examine the manner in which regulations pertaining to the organization and structure of the *polis* entered into what might otherwise be thought of as the private world of family affairs which constituted woman's permanent domain.

4

MARRIAGE AND THE STATE

I

Every Athenian girl expected to be married, and marriage and motherhood were considered the fulfilment of the female role. As Pomeroy has observed:

> The death of a young girl often elicited lamentations specifically over her failure to fulfill her intended role as a wife. Epitaphs express this feeling, and some vases in the shape used to transport water for a prenuptial bath mark the graves of girls who died unwed. The dead maiden is portrayed as a bride on these memorial *loutrophoroi* vases.
>
> (1975: 62)

Yet despite strong sentimental convictions that a woman's happiness lay in marriage and motherhood and that spinsterhood or childlessness were sad fates,[1] the Athenians themselves recognized clearly enough that neither the importance nor the purpose of marriage in their society lay solely in the personal satisfaction it was held to afford women (or, for that matter, men).

It was the social and moral responsibility of a *kyrios* to arrange for the marriage of any girl in his charge; a responsibility which morally, though not legally, also demanded that he dower her.[2] When the speaker of Isaios 10 concludes his address to the court with a vilification of his opponent and an account of his own exemplary character, he declaims:

> But no doubt gentlemen it is not enough forXenainetos to have wasted the *oikos* of Aristomenes by his pederastic pursuits . . . he thinks that he ought to dispose of this *oikos* in the same way.

> I, on the other hand, men of the jury, though my means are small, gave out in marriage my sisters with what I could; and as one who contains himself in an orderly fashion and performs the duties assigned to him and serves in the army, I demand not to be deprived of my mother's paternal estate.
>
> (Isaios 10 [*Aristarchos*]: 25–6; Loeb trans., modified)

Here the speaker lists the 'giving out' and dowering of his sisters as the first example of his moral rectitude. But the allusion has further point. In a case which involved securing a further inheritance, the speaker contrasts the fulfilment of family obligations by having 'given out' his sisters in marriage with his opponent's destruction of family resources by expenditure on another form of sexual union which was the antithesis of family life. And the contrast is not only between personal virtue and personal vice, between a man who is solicitous of his sisters' happiness and a man who is sexually perverted (for homosexuality was not in any simple sense a 'perversion'). Rather, it is a contrast between social virtue and social vice, a contrast between a man who shows his concern for the maintenance of the Athenian *oikoi* by 'giving out' his sisters to fellow Athenian citizens with as much dowry as he can afford, and a man who shows his disregard for the Athenian *oikoi* by excessive indulgence in a form of sexual pleasure that not only wastes his estate but that also fails to regenerate his *oikos* and thereby the *oikoi* of others. With the marrying of women personal obligations towards female kin and social obligations towards the community at large were merged – a point of which we are reminded by the speaker's immediate mention of the performance of public duties and service in the army.

Apollodoros' prosecution of the Corinthian courtesan Neaira is also illustrative of a view of marriage which blends personal obligation with considerations of a more social and public nature. I shall quote at some length:

> And when each of you goes home what will he find to say to his own wife or his daughter or his mother if he has acquitted this woman when the question is asked, 'Where were you?' and you answer, 'Sitting as jury'. 'Trying whom?' it will at once be asked. 'Neaira', you will say, of course, will you not, 'because she, an alien woman, is living with a citizen contrary to the law and because she gave her daughter who had lived by

prostitution to Theogenes, the *basileus*, and this daughter performed on the city's behalf the rites that none may name and was given as wife to Dionysos.' And you will narrate all the other details of the charge showing how well and accurately and in a manner not easily forgotten the accusation covered each point. And the women when they have heard will say, 'Well, what did you do?' and you will say, 'We acquitted her.' At this point the most virtuous of women will be angry with you for having deemed it right that this woman should share in like manner with themselves in what belongs to the *polis* and in the religious rites. And to those who are not women of discretion, you point out clearly that they may do as they please, for they have nothing to fear from you or from the law . . .

Take thought also for the women of citizen birth to see that the daughters of the poor are not left ungiven. As things are, even if a girl is quite without means, the law sees to the provision of a sufficient dowry for her if nature has given her a passably moderate appearance. But if you acquit Neaira and the law has been dragged through the mud by you and made of no effect there is no possible doubt but that because of poverty the prostitute's trade will come to the daughters of citizens, as many as cannot be given out, and the regard we have for free women will be transferred to courtesans if they gain leave to breed children with impunity and to have a share in the rites and religion and honours of the *polis*.

(Demosthenes 59 [*Neaira*]:110–12, 113; Loeb trans., modified)

Much of this highly rhetorical address relates specifically to the assumed feelings and interests of Athenian women: their indignation should Neaira be acquitted; the dire consequences for poor Athenian girls should prostitutes and courtesans be placed in competition with them. But behind this dramatic account of the personal interest which Athenian women may have felt in maintaining their position as the wives, mothers, and daughters of citizens, and behind the emotional appeal to the male jurors to defend that position by convicting Neaira, there lies, as perhaps there always must, a further series of considerations which relate not only to the well-being of women, but to the preservation of a social system. These considerations again involve the conception of the Athenian *polis* as a closed and tightly knit political and religious unity in which the breakdown of regularly conducted marriage would threaten the existence of the community itself (cf. Vernant 1982: 50).

II

But what constituted marriage in Athens? The first problem (though I do not think it a serious one) is terminological. Aristotle remarked in passing that, 'the union (*syzeuxis*) of man and woman is without name' (*Politics*: 1253b 9–10). In other words, Greek appears to have lacked a word for marriage. The word '*gamos*' (vb *gamein*) is often translated as '*marriage*', but in fact it denoted the act of marriage, the physical inception of a marital union, the wedding, and in Homer it is used for both a marriage and a marriage celebration almost interchangeably (including one celebration from which the bride happened to be absent) (Finley 1954: 28–9). What Greek did lack was a *precise* term to designate the enduring state of marriage.[3] In almost all cases the word which was used to cover the situation was '*synoikein*'. Literally, however, the verb '*synoikein*' means no more than 'to live together', 'to share the same house', 'to cohabit'. It does not in itself point towards any particular, well-defined legal, religious, or social institution. It would seem to point towards no more than a *de facto* state of affairs.

Such terminological vagueness need not necessarily present a problem. An institution or an institutionalized relationship depends neither for its existence nor its definition on a term. The literal meaning of '*synoikein*' in no way denies the possibility that in certain or even most contexts it could still refer to a situation quite precisely defined by a series of social and legal regulations. Most modern European languages, for example, do not verbally distinguish between woman and wife. French uses *femme* to cover both, just as German uses *Frau* and modern Greek uses *yineka*, but a distinct social and legal category of wives certainly exists in France, Germany, and modern Greece. In this respect ancient Greek usage was similar. The archaic term *damar* meaning 'wife' or 'lady' occurs, but in classical prose *gyne* denoted equally 'woman and wife'. Of course French, for example, can qualify *femme* by *mariée* should context fail to dispel the ambiguity, and likewise Greek used such phrases as *gyne gamete*, or *gyne engue*te, or even *gyne gamete kata tous nomous* (a woman wed in accordance with the laws) for precision or emphasis. The problem is, however, more than terminological; or rather, terminological vagueness in this case leads to a series of substantive issues in which even the precise meaning of such phrases as *gyne engue*te or *gyne gamete* is, with respect to their social and legal consequences, problematic.

Vernant (1982) is inclined to take the uncertainty of the term *synoikein* at face value. By coupling it with the fact that there was more than one way of living with a woman in Athens, but that in all cases it was mere cohabitation which in the final analysis was constitutive of a marital relationship, he claims that we cannot arrive at an entirely unequivocal definition of Athenian marriage. Vernant believes that even in the classical period the distinction between a 'legitimate wife' (a *gyne gamete*) and a 'concubine' (*pallake*) remained, 'in default of an unequivocal legal definition . . . somewhat hazy and uncertain' (1982: 166–7). With this interpretation I can only disagree and, for reasons which will become clearer later, I would go so far as to claim that when the term *synoikein* was used in the context of certain laws drafted in the fourth century BC, then, despite its broad literal meaning, it referred to cohabitation only with a legitimate wife, a *gyne gamete* or *gyne engue*te. I shall return to this question later.

Vernant's hesitancy is not, however, without reason. Anthropologists such as Leach (1961) and Needham (1971) have argued conclusively that 'marriage' cannot be seen as a cross-cultural and universal category; rather it must be taken as no more than a 'bundle of rights' whose particular composition is specific to each particular society, 'hence all universal definitions of marriage are vain' (Leach 1961: 105). Thus in a society in which a number of forms of sexual cohabitation involving differing 'bundles of rights' were recognized, the application of the title 'marriage' to one of them alone may seem arbitrary. Leach allows, however, 'that the nature of the marriage institution is partially correlated with principles of descent and rules of residence' (1961: 108). In the classical period it is beyond doubt that there was one form of cohabitation in Athens distinguished from all others precisely on the grounds of a correlation with descent: namely, marriage by a form of contract known as '*engue*', to which must also be added the marriages of a special class of women, *epikleroi* or 'heiresses', who were awarded in marriage to their father's closest male kin. I reserve full discussion of *epikleroi* to the following chapter. However, the effect of marriage by *engue* and the marriage of an heiress were the same in this vital respect: only their children were *gnesioi*, that is to say, legitimate. Children of all other unions were *nothoi*, illegitimates. Thus while it may still be somewhat arbitrary to reserve the term 'marriage' for these unions alone, their unique effect in conferring legitimacy is of fundamental importance for an understanding of Athenian social structure and of the place of women within it, and juridically it certainly distinguishes them from other 'less formal' unions. The matter, however, is further complicated by the fact that the relative status of *gnesioi* and *nothoi* also appears to have varied historically, thus altering the value (and no doubt the frequency) of marriage by *engue*. This being so, the task is not only to specify the 'bundle of rights' which constituted Athenian marriage, but also to determine to what extent the bundle of rights deriving from marriage by *engue* was significantly different from those created by less formal unions.

III

In Demosthenes 46 the law on marriage by *engue*, or perhaps a law on legitimacy, is quoted. That this law could be read as defining both a woman married by *engue* and children who were legitimate is in itself significant (Wolff 1944: 75).

> She whom her father or her homopatric brother or her grandfather on her father's side gives by *engue* to be a lawful wife, from her the children shall be legitimate. If there are none of these, if the woman is an *epikleros* (heiress) her *kyrios* shall have her; if she is not an *epikleros*, he shall be her *kyrios* to whom (. . .?) has committed (. . . her? herself?)
>
> (Demosthenes 46 [*Stephanos II*]: 18)

As it stands, the law is by no means clear. The first query must be whether the list 'father, homopatric brother, grandfather on the father's side', exhausts the categories of male relatives capable of giving a woman by *engue*. Plato gives a much more extended list of male relatives who had the right to give a girl in marriage by *engue* in *Laws*: 6774e, but it is dangerous to assume that Plato's proposals mirror the reality of fourth-century Athens. Actual law-court cases, however, provide examples of male relatives more distant than those included in the above list giving away a girl by *engue*. It may be that these cases are covered by the last clause of the law, but again, leaving aside the question of *epikleroi*, this last clause is difficult to interpret. In fact the Greek, *hotoi an epitrepse, touton kyrion einai*, does make grammatical sense, and would read, 'he shall be her *kyrios* to whom she has committed herself.' But this reading appears unlikely, since such a choice being made by the girl herself, and perhaps by a very young girl, would be at odds with every other social and legal provision in Athens pertaining to women. Most likely the subject of the verb has dropped out, and the preferred emendation is to supply 'her father'. If this is correct, then the law is stating that in the absence of a father, homopatric brother, or patrilateral grandfather, the right to give a girl in marriage by *engue* passed to whomsoever the girl's father had named (presumably by will) to act as her *kyrios*. Since such a *kyrios* was in any case likely to have been a collateral relative, it may be that in those speeches in which it appears that a girl has been given by *engue* by a male relative more distant than those listed in the law, this relative's right to do so derived from the fact that he had been appointed *kyrios* by the girl's father, rather than because as a collateral relative he automatically possessed it (though what would happen if the girl's father had failed to appoint a guardian before his death is unknown) (Harrison 1968: 19–21, 109–10).[4]

Further, there are a number of cases in which a woman was given in marriage by *engue* not by one of her male kin at all, but by her previous husband acting as her *kyrios*.[5] The most famous case is that of Demosthenes' mother, whom her husband, on his deathbed, engaged by *engue* to his friend Aphobos.[6] If, however, such a married woman had a father, homopatric brother, or patrilateral grandfather still living, then it might be assumed that any further marriage which her husband and *kyrios* arranged for her in the event of his death or of his divorcing her would have had to be with the consent of those male kin.[7]

Admittedly, the law as stated in Demosthenes 46 is unclear and appears to be incomplete. It does, however, draw attention to the two fundamental principles of marriage by *engue*: first, that women married by *engue* and legitimate children, *gnesioi*, were reciprocally defined categories; second, that for a woman to be married by *engue* she had to be given to her husband by her *kyrios* (whether that *kyrios* was her father, homopatric brother, patrilateral grandfather, or some other guardian in whose official charge she was). In essence the *engue* was not so much a union formed between a man and a woman as an arrangement formed between the male *kyrioi* of two *oikoi* in which the woman was the object of exchange made over for the specific purpose of procreating legitimate children. This is the interpretation which, following Wolff (1944), I shall continue to attempt to substantiate.[8]

IV

It may appear somewhat strange that although the social and legal consequences of marriage by *engue* were of the greatest significance, in that only offspring of such unions were legitimate, from a strictly juristic point the actual form of the contract was ill defined. This may be partially due to the inadequacy of the evidence; though, as Harrison comments, 'Lack of form characterized Athenian contracts in general' (1968: 18n.2). The closest to a formalized expression used at the time of the *engue* comes from Herodotos' account of the giving of Agariste to Megakles by her father, Kleisthenes of Sikyon:

> To Megakles son of Alkmeon I *enguo* my child Agariste according to the laws of the Athenians.
>
> (Herodotos, *History: 6˙130)*

The exchange between father and intended son-in-law in Menander's play, the *Perikeiromene*, perhaps illustrates the fourth-century situation:

> *Pataikos*: I give (*didomi*) you this girl for the ploughing of legitimate children.
> *Polemon*: I take her.
> *Pataikos*: And a dowry of three talents.
> *Polemon*: That too, gladly.
>
> (Menander, *Perikeiromene*: 435ff.)

But, as Harrison comments, 'While some expressions of intention by the two parties to the contract must have been necessary, there is no evidence that any particular form of words was needed' (1968: 18). In fact the formality of the *engue* lay in the legal consequences which derived from it – i.e. the legitimacy of the offspring of the union – and from the public nature of the *engue* in that it was performed before witnesses. Harrison claims that witnesses were not necessary to validate the act but merely to establish, if the need arose, that it had taken place (1968: 18). But I would suggest that to distinguish those aspects of an act which gave it legal validity from those which ensured its public recognition exceeds the subtlety of Athenian law. Further, as Bickerman (1975: 11) has argued, what was at stake between the *kyrios* of a girl and her future husband was not a binding contract, since the husband could always withdraw from it, even after having received the dowry. The sole purpose of witnesses was to ensure the recognition of the progeny of the union as legitimate and therefore heirs to the *oikoi* from which they had descended. It would be most odd if family and friends were not present at the undertaking of the *engue*.[9] Indeed, the absence of witnesses was used in court to argue that a woman had not been given by *engue*. Similarly, although a dowry, a *proix*, generally accompanied a girl on her marriage, it was certainly not a legal requirement on which the validity of the *engue* depended (Harrison 1968: 48–52). There are several cases where men married women by *engue* without a dowry.[10] On the other hand, there is little doubt that the fixing of the dowry was a normal component of the marriage by *engue*, and again, the absence of a dowry could be used in court as at least circumstantial evidence that no marriage by *engue* had taken place.[11]

The form of the *engue* seems, then, rather vague; but what has most exercised scholarly ingenuity is the attempt to determine what the exact legal status of the *engue* was in establishing a 'marriage'. Usually the term *engue* is translated as 'betrothal' or 'engagement' to distinguish it from 'marriage proper' – i.e. the actual *copula carnalis* of the bride and groom. The handing-over of the bride to the groom by her *kyrios* was referred to as the '*ekdosis*', the 'giving out', and *ekdidonai*, 'to give out' was the verb generally employed when a *kyrios* made mention of the fact of his having 'married' a girl to someone. Finally, marriage in the sense of the actual reception of the bride by the groom which would instigate their cohabitation, their *synoikein*, was referred to as the *gamos*, and was generally marked by a wedding celebration, a *gamelia*, in the form of a feast offered to the members of the groom's phratry. The *gamelia* was certainly required by the regulations of some phratries, but again Harrison claims that it 'was not a legal act required to make a marriage valid' (1968: 6 and n.2). Once more it might be suggested that public recognition of an act and its 'legal validity' amounted to the same thing. Thus in Demosthenes 30 the speaker says:

> No man in concluding a transaction of such importance . . . would have acted without witnesses. This is the reason why we celebrate marriages (*gamoi*) and call together those who are closest to us, because we are dealing with no light affairs, but are entrusting the lives of our sisters and daughters, for whom we seek the greatest possible security.
>
> (Demosthenes 30 [*Onetor*]: 21; Loeb trans., modified)

Nevertheless the question arises whether the social and legal consequences of 'legitimate marriage' by *engue* were instigated by the *engue* itself, or whether the further acts of the *ekdosis* and *gamos* were necessary to convert, as it were, a 'betrothal' into a 'full marriage' (Harrison 1968: 6).

Harrison can find no further legal requirements, and it is probable that in most cases the *ekdosis* and the *gamos* would have followed on more or less immediately and automatically from the *engue*. Indeed, the whole problem of the legal status of the *engue* in effecting a marriage would not arise if it were not for the single case of Demosthenes' mother and sister. As already mentioned, on his death-bed Demosthenes' father gave his wife by *engue* to his friend Aphobos; simultaneously he gave his daughter by *engue* to a certain Demophon. Here there had to be a lapse of time between the *engue* and the *gamos* since Demosthenes' mother was not to become Aphobos' wife until after the death of her present husband, and since Demosthenes' sister was a mere infant at the time and was not to be 'taken' by Demophon for another ten years.[12] Further, as matters turned out, neither woman ever married the man to whom she had been contracted by *engue*.[13] It would seem evident, then, that the *engue* was not a binding contract which necessitated the *gamos* and subsequent cohabitation, *synoikein*. As Harrison (1968: 7–8) observes, the only spheres in which the legal consequences of the *engue* as separate from 'full marriage' can be registered are thus in the *kyrieia* or guardianship over the woman's person and dowry. In fact the considerable dowries of Demosthenes' mother and sister were made over to Aphobos and Demophon at the time of their *enguai* and the women were placed in these men's *kyrieia*. Unfortunately little can be concluded from this since the whole of Demosthenes' father's *oikos* was in any case placed under the guardianship of Aphobos and Demophon (and a third, Therippides) until such time as Demosthenes came of age. The control the two men exercised over the women and their dowries could thus have resulted as much from the fact that they were the appointed guardians of Demosthenes' *oikos* as from the fact that the two women had been given to them by *engue*.

The whole affair of Demosthenes' mother and sister and their unconsummated *enguai* seems exceptional, and in all probability the legal consequences would not have been clear to the Athenians themselves. But to ask whether *engue* itself constituted marriage, or whether it amounted merely to 'betrothal' prior to 'full marriage', may also be a quite misdirected line of enquiry. For it begs the very question of what is to be understood by 'marriage' – and to try to determine this by reference to the precise point at which a marriage was legally instigated merely confuses the issue. If it were to be decided arbitrarily that the instigation of sexual relations between a couple and the fact of their cohabitation constituted marriage, then *engue* is not marriage. More to the point, *engue* is not even necessary for marriage since it was quite permissible in Athens, both morally and legally, to cohabit with someone on a more or less permanent basis without *engue*. In this sense, what constituted marriage was *synoikein*, the fact of living together. But if, on the other hand, marriage is defined in terms of descent and the inheritance of certain rights and duties by the offspring of that marriage, then *engue* is marriage, for only the offspring of such a union were legitimate. As Bickerman (1975) has rightly argued, the essential fact of marriage in Athens remained cohabitation, *synoikein*, which may to an extent explain the relative informality of the contract. At the same time, however, legitimacy, which derived from marriage by *engue*, was essential to the social, economic, and political life of Athens. To register this and to see its bearing on the status of women, although it involves some vexed issues of historical interpretation, the relative rights of legitimates (*gnesioi*) and illegitimates (*nothoi*) must be examined.

V

In Demosthenes 43 a law on 'intestate succession' is quoted,[14] of which the last clause reads:

> But no illegitimate son (*nothos*) or illegitimate daughter (*nothe*) is to be a member of *anchisteia* neither with respect to religious

rites (*hiera*) nor secular matters (*hosia*), from the time of the archonship of Euklides.

(Demosthenes 43 [*Makartatos*]: 51)

The *anchisteia*, the 'closest', was a defined group of kin who, among other things, stood to inherit a man's property in fixed order in the absence of any direct descendants. The archonship of Euklides dates the law from 403/2. From this date at least any child of a union not contracted by *engue* was placed firmly outside his parents' kinship group.

Apart from the fact that this law, though reinstituted in 403/2, was supposed to be Solonian in origin (traditional date 594 BC), there is evidence that it, or something very similar to it, was in force earlier in the fifth century. In Aristophanes' *Birds*, performed in 415/14, Peisthetairos disabuses Herakles of the idea that he stood to inherit anything from his father Zeus:

> Why here, I'll read you Solon's law about it: 'A *nothos* is not to be a member of the *anchisteia* if there are legitimate sons (*paides gnesioi*); and if there are no legitimate sons, the goods are to go to the next-of-kin.'

(Aristophanes, *Birds*: 1660 ff.)

It is well to remember that this passage is a joke, and that the law(s) had been tinkered with in some way to produce the comic effect of, 'If A, then no inheritance; but if B, then still no inheritance', thus raising Herakles' hopes only to dash them immediately. Following Wolff (1944: 88f.) I think that the most likely reconstruction is that in the original Solonian law *nothoi* were members of the *anchisteia* and stood to inherit in the absence of any legitimate children, but that at some later date and by the time of the performance of *Birds* the law had been made stricter so that *nothoi* were excluded from the *anchisteia* in all circumstances.[15] At all events, if Herakles as Zeus' illegitimate son was not completely excluded from the *anchisteia* by 415/14, the joke would have no point (Harrison 1968: 67).

There is, on the other hand, a certain body of evidence suggesting that at least in earlier periods of Athens' history the disabilities of *nothoi* had not been so great, and that even in the classical period unions formed otherwise than by *engue* enjoyed a degree of recognition.

Towards the end of his speech in prosecution of the Corinthian courtesan Neaira, which was delivered in 349/8, Apollodoros offers the following definitions:

> For this is what *synoikein* is – to have children and to introduce the sons into one's phratry and into one's deme, and to give out the daughters to men as being one's own. We have *hetairai* for pleasure, *pallakai* to care for our daily bodily needs, and *gynaikes* to bear us legitimate children and to be the faithful guardians of our households.
>
> (Demosthenes 59 [*Neaira*]: 122; Loeb trans., modified)

Three categories of women are adduced: *hetairai*, *pallakai*, and *gynaikes*, which might loosely be translated as 'prostitutes', 'concubines', and 'wives'. What was meant by *hetairai* is fairly clear: they were professional sexual entertainers – though this covered a wide range of women from slave-girls in brothels to free and wealthy courtesans, who formed the only relatively independent class of women to be found in Athens. What was meant here by *gynaikes* seems also clear: they were Athenian women, *astai*, married to citizens by *engue*, whose children were legitimate. What was meant by *pallakai* is more difficult to determine (Vernant 1982: 48).

Throughout his speech *Apollodoros* variously refers to Neaira as a *pallake* and as an *hetaira*. There was obviously no clear separation made in his time, and the distinction he makes in the passages above is perhaps no more than a rhetorical flourish (Vernant 1982: 47–8; Wolff 1944: 74). In fact, a *pallake* seems to have been any woman living in a more or less permanent union with a man, but whose *kyrios* had not given her by *engue*; thus any *hetaira* who took up permanent residence with a man could be referred to as a *pallake*.[16] The question, however, is whether in the classical period there were also *pallakai* who were not *hetairai*, but Athenian women, *astai*, cohabiting with Athenian citizens in unions engaged otherwise than by *engue* and, if so, what status their children had.

In the homicide law quoted in Demosthenes 23 and attributed to Drakon (traditional date 621) it is stated that:

> If a man kills another unintentionally in an athletic contest, or in a fight on the road, or unwittingly in battle, or (caught in adultery) with his wife (*damar*) or mother or sister or daughter or *pallake* whom he keeps for (procreating) free children

> (*eleutheroi paides*), he shall not go into exile for having killed on account of these.
>
> (Demosthenes 23 [*Aristokrates*]: 53; Loeb trans., modified)

Here, the *pallake* is explicitly included in the list of those categories of women who fall under a man's *kyrieia*, and it is justifiable homicide to kill any man found in adultery with her. Furthermore, such a *pallake* is said to be kept for the bearing of 'free children'. These children are not, it should be noted, referred to as *gnesioi*, legitimates; nevertheless it seems clear that both they and their mother were at the time of this law considered as part of a man's *oikos*. It may well have been, then, that in the aristocratic society of the seventh century BC the distinction between a woman married by *engue* and a woman not so married, a *pallake*, was rather less important than in the classical period. The distinction between legitimate and illegitimate children depended more or less on their father's desire to recognize them, the practical value of that recognition again depending very much on the father's status within the aristocratic hierarchy (Harrison 1968: 14–15; Vernant 1982: 48; Wolff 1944: 50–1).

According to Plutarch, in the sixth century BC Solon's legislation exempted *nothoi* from the duty of providing for their aged parents.[17] This legislation would have had little point if previously *nothoi* had not been considered at least partial members of a man's *oikos*. Compare the passage from Aristophanes *Birds*, wherein under Solonian legislation *nothoi* probably had rights of inheritance, at least in the absence of legitimate descendants. To take a broad view, then, it would be tempting to see the marked opposition between *nothoi* and *gnesioi*, and consequently of marriages by *engue* and 'less formal' unions with *pallakai*, as in part the result of Athens' growing democratization from the time of Solon onwards, which placed greater emphasis on the individual citizen's *oikos* and consequently greater strictures on membership of it (Vernant 1982: 50; Wolff 1944: 90).

Even in the fifth and fourth centuries BC, however, there is some evidence of the existence of less formal unions with women who were not merely *hetairai*. Diogenes Laertius gives this account of the situation at the end of the fifth century:

> For they say that the Athenians, on account of the scarcity of men, passed a vote with the view to increasing the citizen

> population, that a man might marry (*gamein*) one Athenian woman but have children by (*paidopoieisthai*) another.
>
> (Diogenes Laertius, *Lives of the Philosophers*: 2.26)

There is little doubt that the other woman referred to here is an Athenian woman, an *aste*, and *paidopoieisthai* was the verb normally used for the procreation of legitimate children. Thus, if the law is genuine, it would appear to allow an Athenian citizen to have children for his household from a woman other than his legitimate wife married by *engue*, that is to say, from a *pallake*.[18] It may also represent a temporary reversion at the end of the Peloponnesian War to an earlier state of affairs (Wolff 1944: 86). But apart from this brief and perhaps exceptional period there is further evidence from the fifth and fourth centuries that suggests the existence of *pallakai* as more or less recognized members of the *oikos*. Even if Drakon's homicide law quoted above dates from the end of the seventh century, it appears that it was not a dead letter in the fourth century. At least an orator can refer to it:

> The law-giver believed so strongly that these things were just in relation to married women (*gynaikes gametai*) that he also enacted the same law in relation to less worthy *pallakai*.
>
> (Lysias 1 [*Eratosthenes*]: 31)

The most conclusive evidence, however, for the existence of women given out by some means other than *engue* comes from Isaios 3, in which the speaker argues that because no dowry was given with the sister of Nikodemos she could not have been married by *engue*. He argues this *a fortiori* from the fact that:

> even those men who give their women as *pallakai* all first come to some agreement about what is to be given to those *pallakai*.
>
> (Isaios 3 [*Pyrrhos*]: 39)

In sum, the status of *nothoi* relative to *gnesioi* appears to have deteriorated during the classical period and the significance of marriage by *engue* to have increased, though even in the fourth century there were probably still some Athenian women placed into households by their *kyrioi* as *pallakai* rather than as legitimate wives.[19] Nevertheless, the importance of marriage by *engue* and the role of the legitimate wife placed into the *oikos* of a citizen by her *kyrios* for the purpose of producing legitimate children is beyond doubt, for, as we have seen, at least from 403/2 onwards, and most probably for the better part of the fifth century, a *nothos* was excluded by law from any rights within the *anchisteia*. This meant that he could not inherit his father's *oikos* in either a material or spiritual sense; nor could he inherit from collateral relatives (which incidentally meant that he could not claim an heiress in marriage); nor could he exercise *kyrieia* over, for example, a legitimate sister (necessarily a half-sister). In short, a *nothos* was excluded from all rights exercised within the domain of private or family religion and law, and from all material and spiritual connections with his father and his father's family. The only exception was his right to receive a *notheia*, an 'illegitimate's portion', from his genitor which, according to different sources, was not to exceed either 1,000 or 500 *drachmai* (Wolff 1944: 75–6).

A question remains, however, about the effect of marriage by *engue* on the possession of rights exercised *outside* the confines of the family; that is to say, the importance of legitimacy within the organization of the state. This is a controversial subject.

VI

The central question in determining the importance to the state of marriage by *engue* is whether or not the rights of citizenship were also dependent on legitimacy. Were *nothoi politai*? – a question characterized at the beginning of this century as a 'long, confused, and unimportant controversy' (Wyse 1904: 280).[20] It is, however, a question of considerable importance, for if citizenship was dependent on legitimacy, then the role of women as the legitimate wives of Athenian citizens and the regulations pertaining to marriage become vital to the very definition of the Athenian *polis*. The exchange of women between the Athenian *oikoi* for the purpose of procreating legitimate children functions not merely as one of the means by which rights of inheritance were transmitted within and between *oikoi* but also as one of the means by which the closed community of *politai* was itself defined. In short, the familial status of women becomes vital to the political status of men.

It must be admitted from the outset that there is no single piece of evidence which will settle the question beyond the shadow of a doubt; but the probability is very high that *nothoi* were not citizens, and this is what I shall argue.

In fact the terms *ta hiera* and *ta hosia* (religious rites and secular matters) employed in the law on intestate succession which specify the disabilities of a *nothos*, are quite comprehensive enough to embrace not only rights of inheritance within the family, but also all rights of citizenship as well, and the terms are frequently used in this manner (Wyse 1904: 535).[21] This is *prima facie* evidence that *nothoi* were not citizens.

Further, in Isaios 8 the speaker says:

> When we were born our father introduced us to the members of his phratry, having sworn an oath in accordance with the established laws that he was introducing [the sons] of a woman who was Athenian and married by *engue* (*ex astes kai enguetes gynaikos*).
>
> (Isaios 8 [*Kiron*]: 19)

It is possible that not all phratries had the same regulations, but other examples of similar oaths can be produced and it seems that in general when a father introduced his son(s) into his phratry he was required to swear that they were legitimate offspring born from a woman married by *engue* (Bonner and Smith 1938: 160). Importantly, the same applied even if the son(s) were adopted. Thus in Isaios 7 the speaker explains that:

> he [Apollodoros] conducted me to the altars and to the members of the *genos* and phratry. These both have the same rule, that when a man introduces his natural son or an adopted son, he must swear with his hands on the victims that the child whom he is introducing, whether his own or adopted, is born of an Athenian woman from a legitimate union.
>
> (Isaios 7 [*Apollodoros*]: 16)

It would seem, then, that phratry membership at least was dependent on a man's being born from a marriage contracted by *engue*, and that *nothoi* were not only excluded from their father's phratry, but also from the possibility of adoption by another citizen and enrolment in his phratry.[22] Consonant with these rules it could be presumed that no *nothos* could himself marry an Athenian woman by *engue*.[23] For it would be very strange were a man who was himself excluded from his natural father's *oikos* and *anchisteia* yet to be able to transmit membership of these to his sons, thus making them the possible heirs to their grandfather.

The problem, however, is what exclusion from the phratries implied for citizenship. After Kleisthenes' reforms of 508/7 the demes became the basic organizational units of the state and, since there existed no central citizen registry, it appears that enrolment on the deme lists constituted official enrolment as a citizen. Prevailing opinion, therefore, has been that *nothoi*, although excluded from all rights of inheritance, material and religious, within the family and *anchisteia*, and although excluded from membership of their father's phratry, were nevertheless citizens and enrolled on the deme lists. In fact we have no evidence to ascertain whether *nothoi* were enrolled on the deme lists or not. For lack of decisive evidence, the accepted view that *nothoi* were indeed citizens has been based on two closely connected assumptions: first, that Athenian society could not have been so inhumane as to deny a man citizenship simply because he was illegitimate; second, that questions of politics and state organization were separate from questions of family organization, kinship, and descent. Both these assumptions have been challenged, and I would uphold that challenge.[24] It remains true, however, that in the absence of conclusive evidence about the enrolment or non-enrolment of *nothoi* on the deme lists, the question of their citizenship or non-citizenship has to be argued on evidence concerned with phratry membership, and this in turn raises the question of the relationship between phratry membership and membership of the *polis*.

There is some evidence that in the fourth century it was possible to be a citizen without being a member of a phratry,[25] but as Andrewes admits, 'it is clear that such a man would be in an uncomfortable and questionable position' (1971: 92). In fact the strongest evidence that phratry membership was not a prerequisite for citizenship is simply Aristotle's failure to mention it in his account of the qualifications necessary for a man's entry to the deme registers (*Constitution of the Athenians*: 42.1). This has the weakness of all arguments *ex silentio*.[26] Further, the view that the phratries were not part of the official organization of the *polis* again rests largely on Aristotle's account of Kleisthenes' reforms (*Constitution of the Athenians*: 21) in which the population of Attica was divided into demes, but in which everyone was allowed 'to retain his connections with his *genos*, his membership of his phratry, and his family rites according to custom'. That Kleisthenes' reforms entailed a new division of the population into demes on a territorial basis (which, it might be noted, immediately reverted to the principle of hereditary membership) is unquestionable; but that the phratries were no longer officially part of the organization of the *polis* but merely private bodies within it seems unwarranted. Indeed, everything would seem to point towards the view that a man's legitimate birth within his father's *oikos* from a mother married by *engue*, his admission into his father's phratry, and his admission into his father's deme as a citizen of the *polis*, were all part of the same process whereby the Athenian state regulated and defined its membership.

In fact Drakon's homicide law (which if genuine is very early, but which was in any case reaffirmed in 409/8) states that in default of certain prescribed relatives (the *anchisteia*) action was to be taken on behalf of any murdered man by the members of his phratry. The assumption is surely that all Athenians were phratry members. But the best proof of the continued importance of the phratries, and indeed of their connection with citizenship, comes from law-court speeches of the classical period itself. In Demosthenes 39 the speaker is claiming sole rights to the use of his name, Mantitheos, which was bestowed on him by his father Mantias. The name was being usurped by his alleged homopatric half-brother, Boiotos. Mantitheos claims that Boiotos was illegitimate, and had been accepted into Mantias' phratry only because Mantias had been tricked by a false oath sworn by Boiotos' mother, Plangon (see p. 37). Boiotos' use of the name 'Mantitheos' is adding insult to injury, since it was the name of Mantias' father, and it was general practice to name eldest sons after their paternal grandfather. Boiotos has thus been flaunting his legitimacy, won by a trick, by taking on his grandfather's name, which Mantias had already bestowed on his undoubtedly legitimate son. This son, the 'real' Mantitheos, now rails against his half-brother in the following manner:

> It is an outrage that whereas thanks to this name [i.e. Boiotos, under which the defendant was registered in Mantias' phratry] you have a share in the *polis* and in the estate left by my father [i.e. Mantias], you should now see fit to fling it aside and take on another name (i.e. 'Mantitheos').
>
> (Demosthenes 39 [*Boiotos* I]: 31; Loeb trans., modified)

Mantitheos then goes on to advise his half-brother,

> cease to make trouble for yourself and cease bringing baseless and malicious charges against me, and be content that you have gained a *polis* and a father.
>
> (ibid., 34)

At the very beginning of his speech Mantitheos gives an account of Boiotos' successful suit, whereby he had maintained his legitimacy:

> [Boiotos] brought a suit against my father [Mantias] and . . . went to court alleging that he was my father's son and that he was being treated outrageously and being robbed of his fatherland (*patris*).
>
> (ibid., 2)

And in a subsequent speech (Demosthenes 40) Mantitheos recounts the proposed compromise between his father, Mantias, and Boiotos' mother, Plangon, whereby Plangon's brothers would adopt her sons and enrol them in *their* phratries:

> For if this was done she said that the defendants [Boiotos and a younger brother] would not be deprived of their *polis*, but that they would no longer be able to make trouble for their father.
>
> (Demosthenes 40 [*Boiotos II*]: 10; Loeb trans., modified)

It seems quite clear from these passages that Mantitheos is assuming that Boiotos' successful maintenance of his legitimacy and his admission into Mantias' phratry amounted to one and the same thing as his securing citizenship; it was membership of a phratry which gave Boiotos 'a share in the *polis*', which 'gained him a *polis*'; had he not managed to maintain his legitimacy and gain admission to a phratry he would have been 'robbed of his fatherland' and he and his younger brother would have been 'deprived of their *polis*'. In other words, legitimacy, which derived from being born of a woman married by *engue*, and which was a prerequisite for phratry membership, conferred on a *gnesios*, a legitimate, not only his rights within the family but indeed his membership of the Athenian state itself.[27]

Those scholars who hold to the view that *nothoi* were citizens and that legitimacy had nothing to do with membership of the *polis* dismiss the above evidence on the grounds that although legitimacy was probably necessary for phratry membership, phratry membership and legitimacy were of importance only for *proving* (as opposed to determining) citizenship. For from the end of the fifth century a marriage by *engue* could be contracted only with a woman who was herself of Athenian birth, an *aste*, and from the middle of the fifth century onwards there is no doubt that Athenian parentage on both sides, the mother's as well as the father's, was a necessary condition of citizenship. Thus, while marriage by *engue* and legitimacy were in themselves irrelevant to citizenship, they were at least indications that a man had been born of an Athenian mother as well as an Athenian father, and it was this which was a necessary qualification for citizenship.

VII

The regulations for the admission of a boy to his father's phratry in fact specify *two* attributes of the candidate's mother to be sworn to by his father: (1) that the woman had been married by *engue*, and (2) that the woman was herself of Athenian birth, an *aste*. Thus in Demosthenes 57 Euxitheos claims that:

> my father himself while he still lived swore the customary oath and introduced me to the members of his phratry, knowing that I was of citizen birth (*astos*) and born of an Athenian woman married to him by *engue*.
>
> (Demosthenes 57 [*Euboulides*]: 54; Loeb trans., modified)

It might be supposed that marriages by *engue* between Athenian citizens and the sisters and daughters of resident aliens or other foreigners or free non-citizens had been fairly common in the earlier periods of Athens' history. The male offspring of such mixed marriages had clearly been citizens, but in 451/0 a law proposed by Perikles excluded from citizenship anyone born of a foreign mother, and a category of mother-foreigners, *metroxenoi*, was created. As Aristotle reports:

> In the archonship of Antidotos, because of the multitude of citizens, Perikles proposed and the law was passed that whoever was not of citizen birth on both sides was not to have a share in the *polis*.
>
> (Aristotle, *Constitution of the Athenians*: 26.4)

Some time later either this law was repealed or it fell into disuse.

At all events it was reinstituted in 403/2 under the archonship of Euklides with the special provision that those who had been born of non-Athenian mothers before 403/2 were not to lose their citizenship, and it remained in force throughout the fourth century. Such *metroxenoi* as these laws created were regularly referred to as *nothoi*, i.e. illegitimates.[28]

Aristotle is the only source that provides an explanation for the law: that it was passed 'because of the multitude of citizens'.[29] Many other explanations have been offered, none of them entirely satisfactory, and indeed Perikles' motivation for proposing the law and the people's motivation for passing it are not of direct concern here. Nevertheless, I would like to suggest that Aristotle's very simple explanation could perhaps be taken at face value. The sheer numerical growth of Athens' population was at odds with the Greek conception of the *polis* as a self-contained community bound together by ties of kinship and religion. It threatened amorphism, and it is perhaps significant that the Athenian Empire, unlike the later Roman Empire, was never capable of absorbing and incorporating its subject populations into itself. The Greek city-state remained incapable of transformation, and the restriction of citizenship to those born of Athenian parents on both sides, although in itself a political innovation, was nevertheless in strict accordance with a mode of social definition based on principles of kinship and religion rather than on purely political considerations in our sense of the word. It might be suggested that Perikles' law and its later reinstitution could be seen as the rear-guard actions of an already anachronistic social formation.

At all events, after 451/0 and again after 403/2 there were two ways in which one could be a *nothos*: either by being born of a union not having been formed by *engue*, or by being born of a woman not herself an Athenian, an *aste*. It is certain that this latter category were not citizens, for so much is clear both from Perikles' law and from its later reinstitution. It should be noted, however, that Perikles' law does not itself call those not born of an Athenian mother *nothoi*; it simply says that they are not to be members of the *polis*. Again this has been used by some scholars to argue that *nothoi* who were not *metroxenoi* were still citizens, otherwise why did not Perikles' law merely state that thenceforth those born of non-Athenian mothers were to be *nothoi*, thus automatically excluding them from citizenship? (Harrison 1968: 65). This view, however, relies heavily on a particular form of wording. In any case the two categories of *nothoi* must soon have coalesced, for after these laws there would have been little point in contracting a marriage by *engue* with a non-Athenian woman, since her children would in any case have been illegitimate, thus frustrating the whole purpose of the institution (Harrison 1968: 62). But if some hesitancy must remain about whether women as *legitimate wives* were important to the state as the transmitters of citizenship, there can be no doubt that from the middle of the fifth century their status as Athenian-born was of fundamental importance – for only Athenian-born women could produce children who were citizens.

VIII

It is now time to return to the question of the meaning of *synoikein*. Some time before 340 the laws relating to marriage with non-Athenians were considerably strengthened. Whereas the laws of 451/0 and 403/2 made the children of non-Athenian mothers into non-citizens, the law quoted in Demosthenes 59 reads:

> If an alien man shall 'live with' (*synoikein*) a citizen woman (*aste*) by any trick or contrivance, he may be indicted by anyone who wishes from amongst the Athenians having the right to do so. And if he be convicted he shall be sold [into slavery] and his property, and the third part shall belong to the one securing the conviction. The same shall hold if an alien woman shall 'live with' (*synoikein*) a citizen, and he who lives with an alien woman so convicted shall be fined a thousand drachmai.
> (Demosthenes 59 [*Neaira*]: 61; Loeb trans., modified)

It is now a heavily punishable offence to *synoikein* with a non-citizen. But what, in this law, is meant by *synoikein*?

Wolff (1944: 66–8) has claimed that in the above law *synoikein* still retained its broad and literal meaning of mere cohabitation, and consequently that the law was making it a punishable offence not only to marry a non-Athenian by *engue* but also to cohabit with any foreign *hetaira* or *pallake*. But Wolff is forced to admit that liaisons with foreign *pallakai* and *hetairai* 'were always most common in Athens, and it is indeed hard to imagine that their radical suppression was either desirable or even possible' (1944: 67).[30] The more reasonable conclusion is that, by the time of the delivery of this speech and the formulation of this law, *synoikein* had become synonymous with the cohabitation of a man with his legitimate wife married to him by *engue*, and that it is to this situation alone that the law refers. Further, in the speech in which the law is quoted, Apollodoros glosses *synoikein* in exactly that manner:

> For this is what *synoikein* is – to have children and to introduce the sons into one's phratry and one's deme and to give-out the daughters to men as being one's own. We have *hetairai* for pleasure, *pallakai* to care for our daily bodily needs, and *gynaikes* to bear us legitimate children and to be the faithful guardians of our households.
>
> (Demosthenes 59 [*Neaira*]: 122; Loeb trans., modified)

As used here, *synoikein* is reserved for cohabitation with a woman who will bear legitimate sons and daughters, i.e. for cohabitation with an Athenian woman married by *engue*.

Wolff denies that the law was exclusively concerned with marriage to a foreigner by *engue* because in this prosecution of Neaira it was never claimed that Neaira had in fact married Stephanos by *engue*, merely that she was a foreign woman, a Corinthian *hetaira*, who was 'living with' (*synoikein*) the Athenian citizen Stephanos. But the speaker, Apollodoros, quite clearly counters this when he says:

> I hear that he [Stephanos] is going to set up some such defence as this – that he is keeping her [Neaira] not as a wife (*gyne*) but as a mistress (*hetaira*).
>
> (ibid., 119)

It follows from this passage that living with a foreign woman as a *hetaira* or *pallake* was not a breach of the law, and that *synoikein* as employed here again refers only to cohabitation subsequent on *engue*. The state, after all, had no interest in the sexual unions of its citizens except in so far as they were productive of children who might claim citizenship – and then it was rigorous. The reason why Apollodoros could prosecute Neaira and Stephanos and claim that they were living together 'contrary to the law' was precisely because Stephanos had introduced a son of Neaira's into his phratry. To have done this he must have sworn, and sworn falsely, that the boy was his own legitimate son. He must have represented Neaira as an Athenian woman married to him by *engue*. Further, and this was one of Apollodoros' major charges, Stephanos had given a daughter of Neaira's in marriage by *engue* to an Athenian citizen – a citizen who subsequently became archon Basileus, which meant that the role of the Basilinna who presided over the religious festivals of the *polis* was being filled by the illegitimate daughter of a Corinthian prostitute. And *Apollodoros* cites the law on this matter:

> If anyone shall give an alien woman in marriage to an Athenian man representing her as being related to himself, he shall lose his civic rights and his property shall be confiscated, and a third part shall go to the man who secures his conviction. And anyone entitled to do so may indict such a person before the *thesmothetai* just as in the case of usurpation of citizenship.
>
> (ibid., 52)

According to this law an alien woman would have been any girl not born of Athenian parentage on both sides – and that would have included the daughter of an Athenian citizen and a foreign *hetaira* or *pallake*. Such a girl would also have been illegitimate, a *nothe*, and it is interesting (though not surprising) to find the law referring to the fraud with the words 'representing her as being related to himself'. By the time of this law the child of a union with a *hetaira* or a *pallake*, (or indeed of any union not formed by *engue*) was not considered as being even 'related' to his or her citizen father.

IX

Though a degree of uncertainty must remain about the precise nature of its consequences, the importance of marriage by *engue* within the social structure of the Athenian *polis* is clear, for on it was based the distinction between legitimate and illegitimate children whose respective rights I have tried to show – rights which, I would argue, included the possession of citizenship itself. But what were the moral and practical implications of marriage by *engue*? Again, in many cases this can best be registered within the context of the law.

The distinction maintained between a legitimate wife, a *gyne gamete*, and other women, *hetairai* or *pallakai*, was in fact a moral as much as a legal one in that strict standards of propriety were attached to the first which did not necessarily apply to the others.

The prostitutes found in the brothels of Athens or hired out as entertainers for the drinking parties of men were held in contempt, though it should be remembered that for the most part they were also slaves. But any woman suspected of promiscuity or infidelity certainly lost her right to the respect that the chaste and legitimate wife of an Athenian citizen commanded. She became the target of moral opprobrium as well as of social and legal discrimination, for her expected behaviour was in sharp contrast to that of a *hetaira*. Thus in Isaios 3 the speaker can argue:

> That the woman whom the defendant has deposed that he gave by *engue* to our uncle was a *hetaira* for whoever wanted her, and not his wife, has been testified to you by the other relatives and neighbours of Pyrrhos who have given evidence of fights, serenades, and frequent scenes of disorder whenever the defendant's sister was with him. Yet no one, I presume, would dare to serenade married women (*gynaikes gametai*), nor do married women accompany their men to feasts in the company of outsiders, especially mere chance-comers.
>
> (Isaios 3 [*Pyrrhos*]: 13–14; Loeb trans., modified)

Indeed, the speaker takes it as axiomatic that if the defendant had given out his sister in marriage by *engue* then she would have been subject to rules of good behaviour, chastity, and fidelity. As it is, he even professes some reluctance to talk about her activities before the court:

> it is clear that her brother gave her out on the same terms to all those who associated with her. And if it was necessary to enumerate them all one by one, it would be no small task. If you order me to do so I shall mention some of them; but if it is as unpleasant to some of you to hear this as it is to me to mention such matters, I shall merely produce the depositions made at the previous trial, none of which they (the opposition) thought fit to contest. Yet when they themselves have admitted that this woman was common to anyone who wanted her, how can it be reasonably conceived that she was a legitimate wife (*gyne engue*te)?
>
> (ibid., 11)

For a woman, even an Athenian woman, being given out by one's brother other than by *engue* could be expected to entail social consequences beyond the illegitimacy of offspring.

On the other hand, the notion of 'living in sin', or the mystical idea that a woman's essentially 'shameful' sexuality can find honourable accommodation only within the confines of legitimate marriage where it is somehow 'sanctified', appear to have had little place in Athenian views on marriage. Almost nothing is heard to the effect that a union with a *pallake* was sinful because it was not contracted by *engue*, or of the notion that it was marriage by *engue* which 'sanctified' a sexual relationship.[31] The price paid for a less formal union was a social one in that children of the union were not *gnesioi*. Religious sanctions applied only to the extent that such illegitimate children were not members of the family cult. The price paid by the woman for such a union was that she could not claim the veneration of being the producer of legitimate children. This being so she was no longer constrained by the rules of chastity which such a position required. What her practical position might then be depended on a strictly personal relationship with a man (and here Perikles' devotion to Aspasia should be remembered). But from a sociological point of view the most that can be said is that she had no clear rights or claims to protection or respect other than those which she could muster by virtue of her own personality. But the possibility for a man of entering into a more or less permanent union even with an Athenian woman otherwise than by *engue*, and the fact that the major disadvantage of such a union lay in the status of children rather than in any *automatic* moral condemnation does not, however, undercut the significance of marriage by *engue*; rather, it underscores it. Instead of being the only recognized means by which a couple could achieve the private satisfactions of conjugality, of *synoikein* in the broad sense of the word, marriage by *engue* was seen for what it was – a means of ensuring the continuity of the *oikos* by the procreation of legitimate children acceptable to the state.

To an extent the practical (as opposed to mystical) nature of the contract of marriage by *engue* is also revealed by the high incidence of remarriage. Though the relatively strict segregation of the sexes and the seclusion of unmarried girls bear witness to the importance placed on premarital female virginity, there was no feeling that a woman's virginity or sexual purity were of such value as to hinder the remarriage of divorced women or widows.[32] In fact divorce, if instigated by the man, was a simple affair, and the remarriage of widows if still of child-bearing age was deemed normal. Thompson (1972) has reviewed the fifth- and fourth-century evidence and finds that there are over fifty cases of remarriage mentioned in our sources, of which thirty are the remarriages of women. The complete acceptability of female remarriage is amply testified, however, by the manner in which Demosthenes' father engaged his wife by *engue* to Aphobos prior to his death, presumably both as a token of respect towards his friend Aphobos and out of a desire to provide for his wife's future security. Similarly the wealthy banker, Pasion, left his wife, Archippe, to his partner, Phormion, by whom she later bore two children. Perhaps a man could show no greater esteem than to bequeath his wife to his friend.

In these situations, however, the moral right of women to enjoy motherhood was not an irrelevant consideration. The speaker of Isaios 2 explains that:

> Menekles approached us with many expressions of praise for our sister and said that he viewed his increasing age and childlessness with apprehension. She ought not, he said, to be rewarded for her virtues by having to grow old with him without bearing children; it was enough that he himself was unfortunate. He therefore begged us to do him the favour of giving her out to someone else with his approval.
>
> (Isaios 2 [*Menekles*]: 7–8; Loeb trans., modified)

And in Isaios 8, the speaker actually adduces as evidence of a nefarious scheme to procure an inheritance the fact that a brother had allowed his sister to continue a marriage with an old man no longer capable of having children:

> For he [the brother] did not try to find another husband for her although she was capable of bearing children to another man.
>
> (Isaios 8 [*Kiron*]: 36; Loeb trans., modified)

Provided that a woman had been a chaste and virtuous wife the fact of previous marriages in no way diminished the possibility of her being contracted into further *oikoi* by her *kyrios* to produce further legitimate children. What men put together they could take apart, and the importance of chastity lay not so much in the idea that for the sake of a woman's own 'purity' she had to remain sexually inviolate as in the idea that her chastity and good behaviour as a wife guaranteed the legitimacy of her children. And though the presumption of a woman's happiness in child-bearing is often mentioned, it was still a woman's ability to produce legitimate children for any *oikos* into which she was contracted that explains the frequency of remarriage. This was the overriding concern. Here, the laws relating to adultery are of interest.

X

If a man was caught *in flagrante delicto* with any woman under a citizen's *kyrieia*, then her *kyrios* had the right to kill the man out of hand. Alternatively, the husband could inflict various bodily humiliations on the adulterer or accept monetary compensation from him, holding him prisoner until he could provide sureties for the payment of the sum agreed. The man so held could later refute the charge by bringing a *graphe* in the courts, but if he lost his sureties were forfeit and the woman's *kyrios* could do with him what he would in the presence of the court, short of wounding him with a sword (Harrison 1968: 33). Such provisions point towards the crime being considered a personal outrage against the woman's *kyrios* for which he could exact personal vengeance. If, however, the husband had not caught the adulterer *in flagrante delicto*, or if he decided not to take matters into his own hands, he could bring one of two charges against the alleged adulterer: either a charge of seduction, or a charge of rape. And here there appears an oddity. If the charge was rape, the penalty was a monetary fine. If the charge was seduction, the penalty was death (Harrison 1968: 34). Dover explains the difference on the grounds that 'rape . . . was not regarded as alienating her [the wife's] affection for her husband and was therefore less of an injury to him than seduction' (1974: 147). Pomeroy maintains that, 'The rapist gained the enmity of the woman and thus posed less of a threat to the husband' (1975: 87).

To an extent these explanations may be correct, and the speaker of Lysias 1 claims that:

> that law-giver prescribed death for adultery . . . because he who achieves his ends by persuasion thereby corrupts the mind as well as the body of the woman . . . gains access to all a man's possessions, and casts doubts on his children's parentage.
>
> (Lysias 1 [*Eratosthenes*]: 33; Loeb trans., modified)

But it is the last phrase of the above statement that points towards the real reason for the difference in penalty between seduction and rape. The law was less concerned with the husband's emotional anguish as a result of his wife's alienated affections (and not much concerned with the wife's own feelings in the case of rape), and more concerned by the threat posed to legitimate descent within that *oikos* (and so ultimately to the composition of the *polis*) by the possible undetected birth of a *nothos* within it. If a woman was willingly seduced and became pregnant, then, as Lacey observes:

> The woman would be compelled to claim that her husband was the father, and his kinship-group and its cult was therefore deeply implicated, since it would be having a non-member foisted upon it, and if she were detected, all her husband's children would have difficulty in proving their rights to citizenship if they were challenged.
>
> (Lacey 1968: 115)

In fact the speaker of Lysias 1 nicely shows just this concern to prevent suspicion falling on the legitimacy of his children. Euphiletos has killed his wife's lover and is pleading justifiable homicide. He recounts his domestic life and the beginning of his wife's seduction to the court:

> Athenians, when I decided to marry and brought a wife to my house, for a while I was inclined not to bother but neither was she to be too free to do as she wished. I kept a watch on her as far as possible and with such observation as was reasonable. But when my son was born I began to trust her and put all my possessions in her hands presuming that this was the greatest proof of intimacy. In the beginning, Athenians, she was the best of wives. She was clever, economical, and kept everything in order in the house. But then my mother died, and her death was the cause of all my troubles. For when my wife attended her funeral she was seen by this man, and, as time passed, he seduced her. He looked out for a slave who goes to market, and, making propositions, corrupted her.
>
> (Lysias 1 [*Eratosthenes*]: 6; Loeb trans., modified)

Euphiletos is careful to draw attention to his wife's virtues *prior* to the birth of his son. All her indiscretion commences *after* that event.

Euphiletos wants to acquit himself of murder; but he does not want to cast doubts on the legitimacy of his heir (Pomeroy 1975: 82).

As for the woman, it is not known what action at law could be taken against an adulteress in the classical period. According to Plutarch, Solon had allowed a father or brother who caught his daughter or sister in *flagrante delicto* to sell her into slavery.[33] But in the fifth and fourth centuries the penalties falling on a woman related to her expulsion from her *oikos* and her exclusion from any further participation in the religious activities of the *polis* (Harrison 1968: 36).[34] In fact according to a law quoted in Demosthenes 59 [*Neaira*]: 87, a husband was legally bound to divorce an adulterous wife on pain of *atimia*, that is to say, the suspension of his own citizenship rights, and as far as can be seen this applied whether the wife had been willingly seduced or raped (Harrison 1968: 36).

In short the result of a woman's adultery was to place her outside the category of women capable of producing *gnesioi*, and her role within the Athenian *polis* and the grounds for her incorporation into its civic and religious life were thus abrogated. She was placed on the same level as a *hetaira* or a foreigner, and the severe penalty of *atimia*, loss of civic rights, which fell on a *husband* who failed to divorce an adulterous wife clearly indicates that the wife's adultery was not seen as (or only as) an injury against the husband. As Lacey puts it, 'The importance of being able to prove legitimacy . . . made adultery a public as well as a private offence' (1968: 113). It should also be noted that the charge of seduction could be instigated by a *graphe*, that is to say it could be instigated not only by the offended husband, but by any Athenian citizen. Harrison's conclusion correctly summarizes the situation: 'the woman and her chastity are hardly protected in their own right, but only because she is the humble but necessary vehicle for carrying on the *oikos*' (Harrison 1968: 38).

XI

Harrison's characterization is, however, put in its clearest light by the rules that governed the control of a woman's person and dowry during and after her marriage, and by the actual terminology employed, to which Wolff (1944) has drawn attention.

In fact the term *engue* (vb *enguan*) was employed in two contexts: first, to denote a marriage contract as discussed above; second, in certain other legal procedures in which it was necessary to put up sureties. The double employment of the term is interesting, for it suggests that similarities might be found in the nature of the two types of transaction. As Wolff describes the situation:

> The active form, *enguan*, was used to denote the action of the *kyrios* who gave the bride in marriage, and of the creditor who accepted the guarantee, while the middle *enguasthai* was the form referring to the groom who received the bride, as well as to the guarantor who undertook to stand surety.
>
> (Wolff 1944: 52)

With reference to the bride the verb was used in the passive. Further, *enguan* appears to have meant literally to 'hand over', and the middle *enguasthai* to have meant to 'receive into one's hands'. As Wolff argues, despite the existence of other terms suitable to denote a conveyance, the same term, *engue*, was used for both a marriage contract and a transaction involving a guarantor, because in both cases something was handed over not for the sake of definite and unconditional acquisition but for a *specific and limited purpose* in such a way as to reserve a certain control for the one who made the transfer. 'The aim of the *engue* was to entrust rather than to alienate the object' (Wolff 1944: 53).

This interpretation of the meaning of *engue* is strengthened if the further usages of the other term employed in the marriage context, *ekdosis* (vb *ekdidonai*), are considered. Literally the verb means 'to give out', but it does not mean to transfer in a definitive sense (for which the verb *apodidosthai* was generally employed). Wolff lists the contractual usages of *edkidonai*: a master putting his slaves at the service of another, a master handing over his slaves to be questioned under torture, a man letting out property on a lease, a master or father entrusting his slave or son to a craftsman as an apprentice, nursing contracts in which a baby is placed in the care of a nurse. In all these contexts there is a common factor:

> someone gives up power over a person or a thing for a specific purpose, and its effect is the transfer of rights in so far as this is required by the purpose. But at the same time it is understood that no definite severance of relationship between transferor and the object will take place. This relationship will

> automatically be restored if and when the purpose . . . for which the *ekdosis* is made ceases to exist
>
> (Wolff 1944: 49)

The general meanings of *engue* and *ekdosis* again point, therefore, towards marriage by *engue* and the giving out of a girl as constituting a limited contract between the *kyrioi* of two households made for the specific purpose of procreating legitimate children. As will be seen, the implications of the terminology are borne out in fact. The choice of a marriage-partner will be discussed at a later stage, but suffice to say that there is little indication that the woman's desires were given much recognition. Further, it might be noted that the conception of marriage by *engue* as a contract formed between two *oikoi* rather than as a compact formed between two individuals (the bride and the groom) is suggested by the frequent occurrence of such phrases as 'to have engaged a woman into an *oikos* worth three talents' rather than to 'have engaged a woman to a man worth three talents'.[35] But marriage by *engue* was a *limited* transfer of rights over a woman made for a *specific* purpose and with automatic reversion to the original controller should that purpose cease to exist. This is most clearly revealed by an examination of the respective rights of the two *oikoi* over the woman's dowry and her progeny.

It is generally accepted that Athenian law regarded the wife as holding the title of her dowry while her husband had only the right to administer it. Harrison's warning (1968: 52), however, that the concept of ownership in Greek thought was extremely fluid, and that the Athenians had no general term to describe the law of ownership should be remembered. The words *echein* (to have) and *kratein* (to have command of) always had a factual, concrete, non-juristic connotation (Harrison 1968: 201). In view of this it is perhaps misleading to talk of the wife possessing the title to her dowry while her husband had only its use. Use was tantamount to ownership. As a rule dowries consisted of money, and if they included other assets a monetary value was placed on them at the time of their payment. Except in the case where the dowry consisted of or included non-fungible assets made over to the husband by way of an *apotimema* (a form of mortgage),[36] there is little doubt that the husband could dispose of his wife's dowry as he wished, and there is no real evidence that his wife's consent was needed (Harrison 1968: 52–4; Schaps 1979: 75–7). But if a husband's control over his wife's dowry can be seen as tantamount to ownership of it, there was one vital respect in which his rights over the dowry were limited: the women he received by *engue* and the rights he exercised over her dowry were inseparable. Control of the dowry could not be held without control (and support) of the woman. Possession of the dowry was conditional on the continued existence of the marital union (Wolff 1944: 61–2).

When, in the event of a divorce, a woman returned automatically to her natal *oikos* her husband was also required to return the amount of her dowry to her original *kyrios*. In cases where the dowry was a considerable sum, a husband would often be required to put up a mortgate (*apotimema*) on property equivalent in value to the dowry he was receiving in order to ensure its repayment if the need arose. Should he fail to repay the dowry he was legally required to pay interest on it at the rate of 18 per cent p.a. to the woman's *kyrios* for her support (Harrison 1968: 55–9).[37] The woman's *kyrios* could sue him for support if he failed to pay this interest. The necessity for the initial evaluation in monetary terms is thus clear. The speaker of Isaios 3 makes the point plainly:

> if a man gives something not valued legally, if the woman leaves the husband, or if the husband sends away the wife, the man cannot get back what he gave away but did not evaluate in the dowry.
>
> (Isaios 3 [*Pyrrhos*]: 35; Loeb trans., modified)

It is worth noting that from this passage it appears that the dowry was repayable irrespective of whether the husband sent away his wife or the wife left her husband. Although there is no direct evidence on the matter it would seem that the dowry was repayable irrespective of the grounds for divorce, that is, even if the husband had sent away his wife because of adultery (Harrison 1968: 55–6).

Similar rules to those operating subsequent to divorce applied if a marriage by *engue* was terminated by the death either of the husband or of the wife. If the husband died, the woman and her dowry returned to her natal *oikos*, from which she would be remarried, if possible with the same dowry. If the wife died, then the husband was again obliged to return his deceased wife's dowry to the *kyrios* of her natal *oikos*. The settlement was different only if there were children. If the wife died, then her husband retained control of her dowry but the dowry was inherited by *her* children.

It was not amalgamated with her husband's property, and a man's children by different marriages inherited their respective mothers' dowries separately from their share of the paternal estate. If the husband died, then the widow had a choice: either she returned to her natal *oikos* taking her dowry with her and leaving her children in her deceased husband's *oikos*, or she elected to remain in her deceased husband's *oikos* with her children under the *kyrieia* of her children's guardian(s). Her children would then inherit her dowry; in the meantime it would be administered by their guardians. When her sons came of age, or if they were already of age, she could place herself under their *kyrieia*, and they would control her dowry, and eventually inherit it – but they were obliged to support their mother (Wolff 1944: 61).

As Harrison remarks: 'This firm statutory provision for the return of the dowry in all cases without exception was undoubtedly a protection of the woman and was generally so regarded' (1968: 56). In Isaios 3 the speaker actually suggests that in cases where a poor girl was being married to a rich man a fictitious dowry might be arranged which the husband would still be obliged to return, thus making it inadvisable for him to discard his wife. Pomeroy (1975: 63) has claimed that the retention of her dowry no doubt improved a woman's chances of remarriage, and certainly one of the purposes of the dowry was to provide a woman with lifelong support. It was a form of maintenance fund set up by her original *kyrios*. If, however, the rules governing the residence and control of a woman after the dissolution of a marriage are considered in conjunction with the provisions for the return of her dowry, then the nature of the marriage contract by *engue* is revealed. The general meanings of *engue* and *ekdosis* make clear that the transfer of a woman from her natal *oikos* into the *oikos* of her husband was never complete, but was conditional on the procreation of legitimate children. Here it should be remembered that a girl's father, and most probably her brothers, retained the right to dissolve her marriage and to demand her and her dowry back. In other words, a woman's natal *oikos* retained administrative rights, as it were, over her child-bearing capacities which they contracted out to other *oikoi* along with a dowry for her support. Marriage by *engue* was a form of exchange between the Athenian *oikoi* whereby women were given out and placed in the households of citizens to bear legitimate children for those households and citizens for the state. But it was an exchange in which the original controllers of a woman, her natal family, retained residual rights: the rights to reclaim and to relocate their women in further households should the purpose of the contract – the procreation of legitimate children – remain unfulfilled, or should the possibility of producing other legitimate children in other households present itself (Wolff 1944: 46–65).

This leaves one important feature of the Athenian marriage system to be explained: the interest that a woman's natal family had in retaining these residual rights and in the procreation of legitimate children within the *oikoi* of other citizens. The answer is to be found in the essentially bilateral nature of Athenian kinship which will be discussed in the following chapter.

5

FAMILY AND PROPERTY

I

One of the more striking features of genealogical material derived from fifth- and fourth-century Attic sources is the frequency of marriage between extremely close kin. Nevertheless, no society that has produced so terrible a drama as *Oidipos Tyrannos* could be considered as having had an entirely complacent attitude towards the question of incest. Indeed, one of Aristotle's criticisms of Plato's proposals for a community of property, children, and wives was that the consequent opacity of kinship relations would result in the possibility of the crimes of assault, homicide, feuds, and slander being committed against the members of one's own family, for:

> All these are unholy if they are committed against father or mother or near relatives as if they were not relatives.
>
> (Aristotle, *Politics*: 1262a; T.A. Sinclair trans., modified)

Aristotle then goes on to include 'incest' among these 'unholy' crimes. Interestingly, however, the context is homosexual:

> It is equally odd that Plato, while making sons to be shared by all, wishes to prohibit sexual intercourse between lovers only, but does not prohibit love nor the other intimacies which between father and son or between brother and brother are most unseemly since even the emotion itself is improper between them. And why prohibit sexual intercourse which is otherwise unobjectionable merely on the grounds of the powerful pleasure it gives, drawing no distinction between that and intercourse between brothers or between father and sons?
>
> (ibid., 1262a)

It is only in his last point of criticism, which deals with the transfer of children from one class to another within the *Republic*, that Aristotle perhaps touches on the possibility of heterosexual incest:

> And such transfers would add greatly to the already mentioned risks – assault, erotic relationships, manslaughter; for those transferred to one of the two lower classes will no longer use the terms brother, child, father, mother, of members of the guardian class, nor will those living amongst the guardian class so speak of the rest of the citizens, which would have set them on their guard against committing such acts because of their kinship.
>
> (ibid., 1262b)

There is, however, little doubt that Aristotle considered such relationships, whether homosexual or heterosexual, to have been both criminal and sacrilegious. Plato's dissolution of the family structure poses for Aristotle, among other things, precisely the problem of incest – albeit with the emphasis not, from our point of view, on its best known forms. And though there appears to have been no specific legal action which could be brought against those who committed incest in classical Athens (Harrison 1968: 22 n.3), certain sexual prohibitions were considered unwritten laws. In Xenophon's *Memorabilia* Sokrates actually refers to them as universal laws:

> Is not the duty of honouring parents another universal law, Hippias? . . . And that parents shall not have sexual intercourse with their children, nor children with their parents?
>
> (Xenophon, *Memorabilia*: 4.4.20; Loeb trans., modified)

In fact, despite Aristotle's censure about practical problems, Plato's own views on the matter were firm. In the *Republic* he specifically rules that men may not have sexual relations with:

> their daughter and mother and the descendants of their daughters and the ascendants of their mother, and similarly women . . . with their son and father and their ascendants and descendants.
>
> (Plato, *Republic*: 461b)

It is, however, in Plato's *Laws* that the Athenian attitude towards incest seems most clearly described:

> *Athenian*: We're aware of course that even nowadays most men,

despite their general disregard for the laws, are very effectively prevented from having sexual intercourse with people whom they find attractive. And they don't refrain reluctantly – they are more than happy to.
Megillos: What circumstances do you have in mind?
Athenian: When it is one's brother or sister whom one finds attractive. And the same law, unwritten though it is, is extremely effective in stopping a man sleeping, secretly or otherwise, with his son or daughter, or making any kind of amorous approach towards them. Most people find not the slightest desire for such intercourse.
Megillos: That's perfectly true.
Athenian: So the desire for this sort of pleasure is stifled by a few words.
Megillos: What do you mean?
Athenian: The doctrine that these acts are absolutely unholy, an abomination in the sight of the gods, and that nothing is more revolting. We refrain from them because we never hear them spoken of in any other way. From the day of our birth each of us encounters a complete unanimity of opinion wherever we go. We find it not only in comedies but in the high seriousness of tragedy too, when we see a Thyestes on the stage, or an Oidipos or a Makaraios, the clandestine lover of his sister. We watch these characters dying promptly by their own hand as a penalty for their crimes.

(Plato, *Laws*: 838b; T. J. Saunders trans., modified)

And it is also worth mentioning that in the *Republic* (571c) Plato illustrates a point by remarking that when a man is asleep and thus freed from the constraints of his reason, in his dreams the animal in him does not draw back even from having intercourse with his mother.

In Athens, then, the idea of incest was capable of arousing horror and disgust and was considered contrary to the law, or at least contrary to those axiomatic principles of moral and social behaviour which were unwritten laws (Harrison 1968: 22). At the same time, however, the categories of kin covered by the incest prohibition were few, and to an extent uncertain. There is no doubt that sexual relations between direct ascendants and descendants were forbidden. Although the passage quoted above from Plato's *Laws* also places brothers and sisters within the incest prohibition, a few lines later the Athenian Stranger advocates the extension of these effective moral deterrents to cover all sexual intercourse other than that between a man and his legitimate wife, and remarks:

> At present, however, the law is effective only against intercourse between parent and child.
>
> (ibid., 839a)

No mention is made of siblings.

In fact there is evidence for a number of cases of a man marrying his half-sister. In two of the cases, however, it is emphasized that the woman was the man's *patrilateral* half-sister.[1] As Harrison (1968: 22) observes, the inference is that marriage with a *matrilateral* half-sister was prohibited while being allowed with a *patrilateral* half-sister, and that *a fortiori* marriage with a full sister was prohibited. Such rules are actually stated, but by late authorities.[2]

II

The extemely limited field of relationships covered by the Athenian incest prohibition is of some importance, for it meant that there were few constraints to stand in the way of close-kin marriages. The absence of constraints does not, however, explain the frequency of such marriages, and it would be tempting to look for positive rules enjoining or prescribing these unions. Harrison states that, 'In principle, the Athenian system was endogamic, as can be seen very clearly in the rules relating to *epikleroi*' (Harrison 1968: 21). In fact the rules governing the marriage of an *epikleros*, an heiress, constitute the only positive marriage rules in Athens, and they will be discussed in due course. For the rest, however, I do not think that there are any 'principles' of Athenian marriage to be found; not, at least, in the sense of actual 'rules', endogamic or otherwise.

It is true, of course, that after Perikles' law of 451/0 and its reinstitution in 403/2 the Athenian *polis* as a whole could be described as endogamic since only children born of Athenian parents on both sides were to be citizens, and marriage by *engue* thus involved the exchange of women between the Athenian *oikoi*. But within the Athenian *polis*, it must be stressed that endogamy was no more than a markedly apparent tendency. It was not systematic or rule-governed. Thompson (1967), for example, has conclusively demonstrated that recorded marriages of first cousins in fifth- and fourth-century sources comply with no particular principle. They were contracted between parallel-cousins and cross-cousins, and between patrilateral cousins and matrilateral cousins. Marriages were also contracted between uncle and niece (again both patrilateral and matrilateral), and between more distantly related relatives.

Despite the frequency of close-kin marriages it appears, then, that Athenian marriage and kinship was, to use Lévi-Strauss' terminology, a 'complex system' in which it was only the limits of marriage which were defined, these limits being 'found in the incest prohibition, which excludes, by the social rule, certain solutions which are nevertheless biologically open' (1969: xxiii). It was a system in which the positive preference for a particular marriage partner could hinge on any number of considerations, which might include whether the desired wife (or husband) was attractive, good-tempered, or belonged to a rich or powerful family. Such a preference 'certainly involves a social criterion but its evalution remains relative, and the system does not define it structurally' (1969: xxiv).

But if the Athenian marriage system was 'complex' in that it did not structurally specify kin with whom marriages were to be contracted, but rather left the choice of a marriage partner to be based on a series of relatively evaluated social criteria, such criteria nevertheless resulted in close-kin marriages. In fact some close-kin marriages might have resulted from romantic attachments. The only Athenian women with whom a citizen was likely to have had any familiarity prior to marriage were those to whom he was already related: cousins, nieces, etc., and inasmuch as there might have been 'love matches' in Athens it could be hazarded that they occurred within the confines of the extended family. But the frequency of close-kin marriages is most readily explicable not in terms of 'romance' but simply as a result of the generalized moral obligation to display loyalty, good-will and assistance towards relatives (cf. Thompson 1967: 281).

An Athenian's first loyalties were towards his kin (Dover 1974: 273). It was thus among his kin that he first sought for a husband for his daughter, sister, or any girl in his *kyrieia*. In Isaios 7, for example, the speaker actually presents to the court as evidence of a family quarrel between uncle and nephew (brother's son) the fact that the uncle had not married one of his daughters to his nephew:

> A convincing proof of their enmity is the fact that though Eupolis had two daughters and was descended from the same ancestors and saw that Apollodoros was possessed of money, yet he gave neither of his daughters to him in marriage.
>
> (Isaios 7 [*Apollodoros*]: 11–12; Loeb trans., modified)

Wealth is one criterion, kinship another.

Equally, women were given in marriage by their *kyrioi* as a sign of good-will to those who were simply friends – a process which, from the Athenian point of view, had the advantage of drawing friends into the privileged circle of relatives. As the speaker of Isaios 2 clearly explains:

> My father . . . was a friend and close acquaintance of Menekles and lived on terms of intimacy with him. There were four of us: two sons and two daughters. After my father's death we married our elder sister when she reached a suitable age. . . . Four or five years later when our younger sister was almost of marriageable age, Menekles lost his first wife. When he had carried out the customary rites over her, he asked for our sister in marriage, reminding us of the friendship which had existed between our father and himself, and of his friendly disposition to ourselves. Knowing that our father would have given her to no one with greater pleasure, we gave her to him in marriage. . . . In this manner, having been formerly his friends we became his relatives.
>
> (Isaios 2 [*Menekles*]: 3–5; Loeb trans., modified)

It should perhaps be noted that the girl's feelings about being married to her father's friend are not mentioned.

Needless to say, such favours could be returned, as Theomnestos relates at the beginning of Demosthenes 59:

> my father . . . himself gave in marriage to Apollodoros, son of Pasion, my sister – and she is the mother of the children of Apollodoros. Inasmuch as Apollodoros acted honourably towards my sister and towards all of us, and considered us in truth his relatives and entitled to share in all he had, I took as a wife his daughter, my own niece.
>
> (Demosthenes 59 [*Neaira*]: 2; Loeb trans., modified)

Here, then, we have something closely resembling an exchange cycle. But it is entirely *ad hoc*. Marriages and repeated marriages in Athens between close kin must be seen merely as part of the operations of an overarching moral code which ideally required the continual extension of favours and preferences towards one's kin.

III

The question remains why the giving of women in marriage should have constituted a favour. In the case of marriages contracted between non-relatives, then a partial answer to the question has already been supplied. Marriages drew friends into the kinship group, joining them by the religious bonds of shared family cults and by the morally sanctioned bonds of social and economic co-operation. Such bonds could then be reforged and reinforced by further marriages. But it was also the case that other advantages accrued to the man who received a wife from her *kyrios*.

It is something of an anthropological orthodoxy to say that women are in themselves social valuables, and this was particularly true in Athens. For though there existed a variety of erotic partners for the wealthier citizen who could afford them, it was only from an Athenian woman given by her *kyrios* by *engue* that a man could supply himself with legitimate children and heirs to ensure the survival of his *oikos* and ancestral cult. This, in terms of Athenian ideology, was imperative. Whether *astai*, Athenian women, were in excess or short supply cannot be determined,[3] but the role of the *aste* as sole supplier of legitimate children gave her a unique value, and Apollodoros' rousing words on the subject should be recalled. It is, however, in the economic sphere that the benefits of marriage are most readily appreciable. For the poor citizen, and especially for the peasant farmer, the possession of a wife was probably a necessity in terms of the free labour she provided; but for the wealthier classes (to which most of the evidence relates) the economic advantages of marriage were much more immediate.

First, of course, there was the dowry, the *proix*. As has been shown, the dowry was effectively in the ownership of the husband, and there is no doubt that the dowry was considered in part to be a contribution from the bride's *oikos* to the *oikos* of her husband. A girl's *kyrios* was under strong moral obligation to dower her and to dower her with as much as possible, both in order to find her the best husband (preferably a wealthy husband), and as an expression of his own standing (Schaps 1979: 74–5). A woman's dowry could constitute a considerable fortune, and the speaker of Isaios 3 states that no adopted son would dare give away the legitimate daughter of his adopted father with less than one tenth of the paternal estate (though it may be that an adopted son would feel himself under some special obligation to appear generous). Though the available evidence is slight, Schaps (1979: 78) concludes that the dowries of women from wealthy families ranged from 20 per cent of the paternal estate to less than 5 per cent. As Schaps notes, however, there is no reason to expect that a father's wealth and a girl's dowry would have been strictly proportional. Any number of factors could affect the amount. The most that can safely be said is that a wealthy man was expected to dower his daughter well, and that conversely a large dowry might be expected to facilitate a marriage with a wealthy man. For the poorer classes there is no real evidence but, as Schaps again suggests, on the presumption that the same moral obligation to dower as well as possible existed, then we might suspect that the dowries of the poor constituted a greater rather than a lesser proportion of the family's wealth (1979: 79). At all events, the giving of a woman in marriage, whether to a friend or to a relative, involved an immediate transfer of wealth to the woman's husband, and this, there is every reason to believe, was a substantial consideration in the formation of a union.[4]

Second, and just as importantly, there was always the strong possibility that far more than a dowry might accrue to a man and his descendants as a result of marriage. Here the structure of Athenian kinship and the laws of Athenian succession must be considered in some detail.

IV

The Athenian kinship system appears to be readily recognizable as cognatic. At least the kinship terminology is thoroughly bilateral: ego-centred, descriptive, and with no terminological distinctions made between matrilateral and patrilateral relatives, e.g. between a mother's brother and a father's brother, both of whom were called *theois*, 'uncle', or between a mother's brother's son and a father's brother's son, both of whom were called *anepsios*, 'cousin'. In the classical period a single term, *kedestes*, denoted any relative by marriage, any affine (Thompson 1971). Little, however, can be deduced about kinship structure from a terminology. The manner in which rights of one sort or another were transmitted is a more important index. But here again the Athenian system appears to have been basically bilateral.

The term *'oikeioi'*, derived from *'oikos'* (the house, household, family) and which I have thus far translated as 'relatives', was loosely and generally used to refer to all members of a bilateral kinship grouping. It was not in any sense a well-defined or technical term, and the boundaries of its application seem to have been contingent on the actual familiarity of a man with his kin. It was used by him to refer to all those with whom he lived in close and constant contact and who entered into the general spirit of co-operation and affection which ideally bound all kinsmen. It also marked those who shared in the same family cults, the cults of their interrelated *oikoi* (in the restricted sense of a man and his direct ascendants and descendants). Such *oikeioi* certainly included affines.

In this context it is important to note the relationship which existed between a grandfather and his *daughter*'s sons, his *thugatridoi*, for although they belonged to the *oikos* of their father and not to the *oikos* from which their mother had come, they were nevertheless counted among their maternal grandfather's *oikeioi* and had a share in his ancestral cult. Thus in Isaios 8, the speaker says:

> For, as was natural, seeing that we were the sons of his own daughter, *Kiron* never offered a sacrifice without our presence. Whether he was performing a great or a small sacrifice we were always there and took part in the ceremony. And not only were we invited to such rites, but he always took us into the country for the Dionysia and we always went with him to public spectacles and sat at his side, and we went to his house to keep all the festivals. And when he sacrificed to Household Zeus – a festival to which he attached special importance and to which he admitted neither slaves nor free men outside his family, and at which he personally performed all the rites – we participated in this celebration and laid our hands with his upon the victims and placed offerings side by side with his, and took part in the other rites, and he prayed for our health and wealth, as he naturally would, being our grandfather.
>
> (Isaios 8 [*Kiron*]:15–17; Loeb trans., modified)

Further, though the common practice was to name a son after his paternal grandfather (thus resulting in the same name alternating down a line of direct male descendants), it was also quite common to name a son, usually a younger son, after his maternal grandfather – a practice which could lead to two cross-cousins both bearing the same name, the one named after his father's father, the other named after his mother's father.

The most important manifestation of the bilateral nature of the Athenian kinship system is, however, to be found in the constitution of the *anchisteia*, a grouping already mentioned.[5] Every Athenian citizen stood at the centre of such a legally defined bilateral group of relatives. Most of information about the constitution of the *anchisteia* comes from the law of intestate succession preserved in Demosthenes 43 and paraphrased in Isaios 11 (two closely interconnected law-suits concerned with rival claims to the estate of a certain Hagnias). Unfortunately neither text is reliable. There is a serious lacuna in the first, and the second is but a paraphrase. The first query is the exact extent of the *anchisteia*. It is said to extend 'down to the children of cousins'. It is, however, quite unclear as to whether the law was thereby referring to the children of *ego*'s cousins (i.e. to *ego*'s first cousins once removed) or to those children who, along with *ego*, had fathers and mothers who were all cousins (i.e. to *ego*'s second cousins).[6] At all events, the *anchisteia* was a bilaterally constituted grouping of collateral relatives and in law it had two functions. First:

> For the law of homicide it determined the body of relatives of a slain man on whom, in the absence of direct descendants, was laid the responsibility of avenging the dead man and whose unanimous consent was needed if the slayer was to be freed by an act of composition.
>
> (Harrison 1968: 143)

Second, and more importantly, it described the body of relatives who *in the absence of direct descendants, natural or adopted*, had the right to claim, in fixed order, the property of the deceased. Its net effect was to determine the limit of the relatives of the deceased's father who could claim the deceased's estate before relatives on his mother's side became entitled (Harrison 1968: 143).

In Demosthenes 43 the law of intestate succession relating to the *anchisteia* reads:

> Whoever dies without having bequeathed [his property], if he leaves female children [the property] is with them; if not, the following are to be in control of his property: If there are brothers [of the deceased] by the same father [they are to have the property]. And if there are legitimate sons of these brothers, they shall take the share of their father. If there are no brothers, or sons of brothers . . . their descendants are to share in a like manner. Males and the children of males are to take precedence if they are of the same stock, even if they are more remote kin. If there are none of the father's side down to the sons of cousins, those on the mother's side are to be in control in the same fashion. But if there are none on either side within this degree, he who is closest on the father's side shall control [the property]. No male or female illegitimate shall be a member of the *anchisteia* neither with respect to religious or to civic rights, from the time of the archonship of Euklides.
>
> (Demosthenes 43 [*Makartatos*]: 51)

While in Isaios 11 the paraphrase reads:

> the law has given the property of a brother first to his brothers and his nephews, providing that they are of the same father, for they bear the closest relationship to the deceased. If there are none of these, the law names secondly his sisters by the same father and their children. If there are none of these, it gives the rights of next of kin to the third degree, to cousins on the father's side down to the sons of cousins. If they are lacking also, the law goes back and makes those on the mother's side controllers of the property in just the same fashion as it originally gave the property to those on the father's side.
>
> (Isaios 11 [*Hagnias]*: 1–2)

To an extent the lacuna in the first text can be supplemented by reference to the second. The law must at least have contained the words, 'sisters by the same father and their children'. The longer emendation, 'sisters by the same father and their children are to share (the property). If there are no sisters or children of sisters, the brothers of (the deceased's) father, and the sisters, and their children (are to share in a like manner)' has also been proposed (Harrison 1968: 144 n.2; cf. Wyse 1904: 680–1). Further, knowledge of the order of intestate succession can be supplemented by reference to the claims and counter-claims made in actual law-court suits. It must be stressed, however, that the order is by no means sure, since it is difficult to assess the veracity with which speakers were interpreting the law and there are numerous inconsistencies. In all probability, the exact application of the laws of inheritance and succession were uncertain even in fifth- and fourth century Athens and left ample scope for *ad hoc* interpretation.

The various problems arising from the laws of inheritance and succession are too numerous and too complex to be pursued here, but following Harrison (1968: 144–9) the most probable order of succession within the *anchisteia* was as follows:

(1) Brothers of the deceased and their descendants without limit.
(2) Sisters of the deceased and their descendants without limit.
(3) Paternal uncles, their children and grandchildren.
(4) Paternal aunts, their children and grandchildren (possibly followed by (4a) paternal great-uncles with their children and grandchildren and (4b) paternal great-aunts with their children and grandchildren – thus extending the *anchisteia* down to the deceased's second cousins).
(5) Brothers of the deceased by the same mother and their descendants without limit.
(6) Sisters of the deceased by the same mother and their descendants without limit.
(7) Maternal uncles, their children and grandchildren.
(8) Maternal aunts, their children and grandchildren (possibly followed by (8a) maternal great-uncles with their children and grandchildren and (8b) maternal great-aunts with their children and grandchildren – again extending the *anchisteia* to second cousins). If there were no relatives within the *anchisteia* then the property simply passed to the nearest relative on the father's side.

It can readily be seen that there is a patrilateral bias in the order of succession. As already noted, the *anchisteia* has the effect of setting the limit to the patrilateral relatives who can inherit before matrilateral relatives become entitled. But even given this initial bifurcation of the *anchisteia* into a patrilateral grouping (claims 1–4) which had precedence over the matrilateral grouping (claims 5–8), rights of succession vested in women or rights transmitted through connections with women were among the first which could be exercised.

Thus, for example, the claims of homopatric sisters ranked second only after the claims of homopatric brothers. To take an actual case, in Isaios 3 the speaker presses the claim of his mother, the sister of Pyrrhos, to Pyrrhos' estate. Alternatively a man could claim an estate on his own behalf by virtue of the fact that his deceased mother had been the sister of the man whose estate was now claimable, i.e. he could claim the estate of his maternal uncle. Again this claim ranked second only after the claims of the deceased's brothers and their descendants. In other words, the effect of the patrilateral bias was only to make an *initial* distinction between claims of relatives on the deceased's father's side and claims of relatives on the deceased's mother's side. But within the patrilateral half of the *anchisteia* claims in the name of a woman or claims through a woman came quickly into play.

This situation was modified only by the covering clause 'males and the children of males are to take precedence, if they are of the same stock (as females and the children of females), even if they are more remote kin'. The simplest interpretation of this clause is that, for example, the son of a homopatric brother would exclude a homopatric sister, for though they were both of the same stock, i.e. had a common ancestor, the male took precedence even though being more distantly related to the deceased, i.e. being a nephew as opposed to a sibling (Harrison 1968: 147).

But since women were not legally competent to control any property, despite being nominal claimants to estates, it was their *kyrioi* who stood to benefit from any female inheritance (cf. Schaps 1979:24). Here one of the financial benefits of marriage can be seen, for a man might find himself the effective beneficiary of an estate which, for example, his wife had inherited on the death of her childless brother. And it should be noted that such advantages were there to be gained for a man *already* closely related to the woman he was marrying. For in many cases it was less a question of a man marrying into a wealthy family as an outsider, than a man moving *closer* within the family to a source of wealth. To take a hypothetical example: a man X marries the sister Y of a childless relative Z; X is in fact the paternal uncle of the siblings Y and Z, i.e. X has married his brother's daughter. Now when his childless nephew Z dies without any direct heirs, natural or adopted, X does in fact have a claim on Z's property in his own right, as Z's father's brother. But this claim ranks only third within the *anchisteia* (and

Z may have had other paternal uncles). However, his wife and niece has second claim to the property, and providing she has no other siblings, she would stand to inherit her brother's property. Thus by marrying his niece, a man has improved his effective claim to a property from third (as paternal uncle) to second (as *kyrios* of the deceased's sister), and though of course no man could count on his nephew dying before him, he could at least ensure that his children would have a better claim on the deceased's property. For whereas before they would have been the deceased's first cousins (father's brother's children), now they are the deceased's nephews or nieces (sister's children), and though their claims would be through their mother, such claims would still be the second-ranking claims within the patrilateral half of the *anchisteia*.

V

Despite the essentially bilateral nature of the *anchisteia* which I have just emphasized, there were nevertheless important respects in which the Athenian kinship system can be seen as patrilineal. Thus far I have been concerned with the consequences of so-called intestate succession. If the more normal circumstances for the transmission of an *oikos* are examined a strongly patrilineal element can be seen at work both in the rules which governed its transmission and in the ideology on which those rules were founded. This substantially modifies or complicates what has been described above. Importantly, however, even here the role of women comes into play, and there is an interesting accommodation made between bilateral and patrilineal succession.

If a man had direct male descendants, natural or adopted, then they inherited his *oikos* in both the material and the spiritual sense, automatically, necessarily, by equal division, and to the complete exclusion of any daughters and, of course, of any collateral relatives (i.e. the members of the *anchisteia*). No daughters or descendants of daughters had any claims in the presence of direct male heirs. Even their dowries were not an inheritance in the strict sense of the word, for if their father had died then the amount with which they were dowered depended on the good-will and self-esteem of their brothers who, legally, shared the whole of the patrimony between them. Nor, in the presence of legitimate sons, could a man in principle bequeath his property in whole or in part to anyone else, whether relatives or friends. Equal division between direct male descendants was automatic and sanctioned not only by morality but by law (Harrison 1968: 48–9, 130–2, 151). Thus in Isaios 10 the speaker explains:

> If, therefore, anyone shall assert that Aristarchos made over his property, he will not be speaking the truth; for while he possessed a legitimate son, Demochares, he could not have wished to do so, nor was it possible for him to give his property to anyone esle.
>
> (Isaios 10 [*Aristarchos*]: 9; Loeb trans., modiified)

Further, if a man did not have legitimate sons, natural or adopted, then although collateral relatives within the *anchisteia* could inherit his property as described above, his own *oikos* would itself die out. He would have no one to carry on his ancestral cult and no one to perform the customary rites and sacrifices at his ancestral altars both for himself and for his forefathers. Nor could any daughter supply the desired continuity of the *oikos* in her own right. Without sons, a man's property would most certainly find a claimant, but his *oikos* would cease to exist, and in terms of Athenian ideology and religious belief this was a catastrophe. Indeed, 'that an *oikos* should not be made empty' was one of the common-place pleas of Athenian law-court rhetoric, and there is every reason to believe in its emotional force.[7]

The Athenian kinship system was thus insistent on the necessity of finding a direct male heir for an *oikos*, not to inherit its property (for with the rules of bilateral inheritance that presented no problem), but for its continued existence (one might almost say its metaphysical existence) as an *oikos*. In this sense, the Athenian kinship system was strongly patrilineal.

Such an ideology of strict patrilineal succession does, of course, create difficulties, for a man may find himself approaching death with no sons to continue his *oikos*. This situation appears to have been not uncommon in Athens (at least it is not uncommon in our sources). The problem was solved by the expedient of adoption, of *eispoiesis* (literally, 'making in'). As the speaker of Isaios 7 explains:

> All men when they are nearing their end take precautions on their own behalf to prevent their *oikoi* from becoming extinct and to ensure that there will be someone to perform sacrifices and to carry out the customary rites over them. And so, even

> if they are childless, they adopt children and leave them behind them. And there is not merely a personal feeling in favour of this course, but the state has taken public measures to endorse this since the law entrusts the archon with the duty of preventing *oikoi* from becoming extinguished.[8]
>
> (Isaios 7 [*Apollodoros*]: 30; Loeb trans., modified)

Solon's law on testamentary succession, which enabled adoption, is cited on numerous occasions.[9] In Isaios 3, for example, we read:

> For the law expressly states that a man is allowed to dispose of what is his to whosoever he wishes, provided that he does not leave legitimate male children.
>
> (Isaios 3 [*Pyrrhos*]: 68; Loeb trans., modified)

In fact the right to make such a testament amounted to no more than the right to *adopt* a son, whether *inter vivos* or by will, who would then inherit the *oikos* in full, claiming its property, but also perpetuating its existence (Gernet 1955: 122ff.; cf. Harrison 1968: 149–50). As the speaker of Isaios 9 explains:

> It is only reasonable to suppose that Astyphilos did not only feel the desire to adopt a son, but also provided that whatever dispositions he made should be as effectual as possible, and that whomsoever he adopted should both possess his wealth and have access to his ancestral altars and perform all the customary rites for him after his death and for his forefathers.
>
> (Isaios 9 [*Astyphilos*]: 7; Loeb trans., modified)

Any adopted son was required to renounce all rights of succession within his own natal *oikos*, and no man could become the simultaneous heir to two *oikoi*. The right of a childless man to adopt a son from another *oikos* (usually, of course, from an *oikos* which had produced a number of sons) was thus designed to maintain the number of the Athenian *oikoi*. But in thereby allowing a man to determine by an act of volition to whom his property should fall, it must also be seen as a considerable abrogation of the rights of collateral relatives within the *anchisteia*, who otherwise would have stood to inherit their deceased relative's property but without continuing his *oikos* (Gernet 1955: 121ff. and 1920: 123ff.). In fact, the right to adopt an heir can be seen to be in accord with the generally 'democratic' tenor of Solonian legislation with its emphasis on the rights of the individual Athenian citizen. A citizen's claim to perpetuate his own *oikos* even if he died childless becomes rated higher than the claims of his collateral kin to absorb his property. But if this Solonian legislation can be seen to be emphasizing a certain individualism at the expense of the claims of kinship, it can also be seen to entail a shift from a bilateral kinship system to a patrilineal one. The individual's right to dispose of his property is precisely a right to ensure the existence of his own *oikos* down a direct line of male descendants by the expedient of adoption.

In this situation we might expect to find some opposition between relatives who stood to inherit by virtue of their position within the *anchisteia* and sons who had been adopted by childless men to become their heirs. In fact, many of the recorded inheritance disputes take just this form. Further, there was one important respect in which an adopted son's rights to his adopted father's *oikos* were not identical with those of a natural son: no adopted son could in turn adopt an heir. If an adopted son failed to procreate heirs for his adopted *oikos* then the rights of the *anchisteia* to his adopted father's property came automatically into play, and Gernet (1955: 128) is willing to see this as a conservative concession to the rights of bilateral kin. At this point, however, an interesting fact should be noted that shows the accommodation of the patrilineal ideology of the *oikos*, now fully upheld by law, to the more general ideology of loyalty to be shown towards one's kin who were still defined bilaterally.

Although the law stated quite clearly that a childless man could adopt whomsoever he pleased to be the heir to his *oikos*, in practice childless Athenians did not adopt complete outsiders but generally adopted a member of their own bilateral kin. Thus, for example, in Isaios 7 the speaker, Thrasyllos, claims to have been adopted by his mother's homometric half-brother, Apollodoros, but finds his claim to Apollodoros' estate challenged in the name of the late Apollodoros' patrilateral female cousin (father's brother's daughter) who stood closer within the *anchisteia*. In defending the fact of his adoption, Thrasyllos argues first that there had been a family rift between Apollodoros and his patrilateral kin, and then goes on to say:

> Since such was the disposition of the cousins towards one another, and so grave the enmity towards Apollodoros who

> adopted me, how could he have done better than to follow the course which we did? Would he, by Zeus, have done better to have chosen a child from the family of one of his friends and adopted him and given him his property? But even such a child's own parents would not have known, owing to his youth, whether he would turn out a good man or worthless. On the other hand, he had experience of me, having sufficiently tested me. He well knew what had been my behaviour towards my father and mother, my care for my relatives and my capacity for managing my own affairs.
>
> Further, I was no outsider, but his own nephew, the son of a half-sister.
>
> (Isaios 7 [*Apollodoros*]: 33–4; Loeb trans., modified)

Thrasyllos admits that it would have been perfectly possible for Apollodoros to have adopted the son of a mere friend, but his assumption is that it would be far more reasonable to adopt a relative. Since Apollodoros was at odds with his patrilateral relatives, to adopt the son of a half-sister was an obvious choice – a half-sister by his mother, not his father. Similarly in Isaios 2 the speaker, though stressing a man's right to adopt whomsoever he wished, assumes that a man who desired to adopt a son would not travel outside the circle of his relatives (Lacey 1968: 145):

> I would dearly like my opponent, who thinks himself so clever, to tell me which of his relatives (*syngenoi*) Menekles ought to have adopted? . . . Well, then, the son of his sister or of his male or female cousin? But he had no such relative at all. He was therefore obliged to adopt someone else, or, failing that, to grow old in childlessness as my opponent now thinks he ought to have done. I think, therefore, that you would all admit that when he adopted a son, he could not have adopted anyone who was more closely connected with him (*oikeioteros*) than I was. Otherwise, let my opponent indicate such a person. He could not do so, for he had no other relative than those whom I have mentioned.
>
> (Isaios 2 [*Menekles*]: 21–2; Loeb trans., modified)

It is important to note in the above that while the speaker was not a *syngenes*, that is, a relative by common descent, he does claim to have been his adopted father's *oikeioteros* relative, the relative 'most closely connected in the family, the *oikos*'. For in fact the speaker was the brother of Menekles' sometime wife.

In short, the rule demanding the supply of a direct male heir for the continuation of an *oikos* (and an ideology which made that continuation imperative) is combined with the overriding moral obligation to extend favours towards kin in such a way as often to result in the transformation of *matrilateral* or *affinal* relatives into direct *patrilineal* descendants by the process of adoption.[10] Despite the strictly patrilineal succession required for the maintenance of an *oikos*, as a result of adoption relatives connected by marriage or through maternal links still stood to become the heirs to a vacant *oikos*. By marrying his sister to Menekles, the speaker of the above passage had placed himself in the likely position of becoming his brother-in-law's adopted son and of inheriting his wealth. Similarly, in Demosthenes 41, a certain Polyeuktos, who had no male children but did possess two daughters, adopted Leokrates, his wife's brother, as heir to his *oikos*, at the same time giving him one of his daughters in marriage. Thus in a sense Leokrates was simultaneously Polyeuktos' brother-in-law, son-in-law, and son. Once again, by strategic (or fortuitous) marriages a man could place himself in the position of inheriting a considerable fortune, or of securing one for his offspring.

It is in this accommodation between the rules governing the patrilineal succession of the *oikos* and the essentially bilateral nature of Athenian kinship that certain specific features of Athenian marriage can best be understood.

In the previous chapter it was noted that a woman's natal *oikos* retained a strong interest in her and in her dowry even after marriage, and that it could be seen as retaining residual rights over the contracting out of a woman's reproductive abilities to bear legitimate children. We are now in a position to answer the question of why (sentimental attachments aside) this was so. Although by the rules of patrilineal succession neither a woman herself nor her sons could automatically supply her natal *oikos* with a direct heir, a woman could nevertheless produce sons who, by the process of adoption, most certainly could become heirs to her natal *oikos*, and who in fact tended to be favoured choices. Thus the legitimate children that a woman bore even within another *oikos* created, as it were, an alternative supply of heirs for her natal *oikos* should the need arise. By the process of adoption, a sister could supply her brother with a son. Further, a girl could supply her own father with a son (Wolff 1944: 50), which brings us to the subject of *epikleroi*.

VI

The rules governing *epikleroi* are intriguing and, owing to the nature of the evidence, it is difficult to determine their full effect. I shall be attempting to give here no more than a broad outline of the subject.[11]

As we have seen, if a man had legitimate sons, natural or adopted, then they automatically inherited their father's *oikos* to the complete exclusion of all other relatives, including daughters. If a man had no natural children, and he failed to adopt a son, then his *oikos* would become extinct and his property would be inherited by his next-of-kin within the *anchisteia*. A woman could not in her own right supply the desired continuity of an *oikos*. So if a man had a daughter or daughters but no son, he still lacked an heir for his *oikos*. His daughter, however, was said to be *epikleros*, that is, 'with the property', a term which is usually, though misleadingly, translated as 'heiress'.

A man with only a daughter, an *epikleros*, could still adopt a son in just the same manner as a childless man, but with one important *proviso* – that any son he adopted was required simultaneously to marry the *epikleros*. This provision illustrates the fundamental feature of all rules governing the *epikleros*: that she is 'with the property' and that no man could inherit the property and become its *kyrios* without at the same time becoming her husband and *kyrios* (Lacey 1968: 140).

The rules governing the *kyrieia* and marriage of an *epikleros* are thus broadly speaking parallel to those governing the inheritance of an estate, for the *epikleros* was really an attachment to the estate rather than an heiress. A man could bequeath his property and his *oikos* to whomsoever he wished, provided that he married his daughter to that man, who, of course, also became his adopted son and direct heir. If, however, he failed to bequeath his property by adopting the man to whom he married his daughter, or if he died before arranging the marriage of his daughter, then the fate of the *epikleros* was practically the same as the fate of the property in the case of intestate succession: i.e. the daughter could be claimed in marriage along with the property to which she was attached by her father's closest male relative within the *anchisteia* (Harrison 1968: 132–3, 151).

There is slight evidence that the next-of-kin of an *epikleros'* father was under some obligation to marry her rather than simply having the right to claim her in marriage along with the property.[12] However, he could certainly decline to marry her, in which case she could be claimed by the next in line within the *anchisteia*. The order of claimants for the hand of an *epikleros* and for her father's property was presumably the same as for male claimants in the case of intestate succession: the first claimant would be her father's brother, the next her father's brother's son, etc. At all events, at such time as a girl became *epikleros*, that is, on the death of her father or possibly on the death of a brother which thus left her 'heiress' to her father's estate, she became *epidikos*, 'adjudicable', and was awarded to the best qualified claimant amongst her father's kin by a process of *epidikasia*, 'adjudication', under the supervision of the archon. If there were rival claimants the issue was settled by a court sitting under the archon as president (Harrison 1968: 9–11; Wyse 1904: 348–9).

The point which perhaps strikes the modern reader as most bizarre is that it is quite clear from the statements of the speaker of Isaios 3 that a woman who became *epikleros* could be claimed in marriage by her father's closest male relative, along with the property to which she was attached, even if she was already married to somebody else (on the assumption, of course, that her father had failed to adopt her husband as his son and heir):

> The law ordains that daughters who have been given in marriage by their fathers and are living with their husbands – and who can judge better than a father what is in his daughter's interest? – in spite of the fact that they are thus married shall, if their father dies without leaving them legitimate brothers, pass into the legal control of their next-of-kin. Indeed, it has frequently happened that husbands have thus been deprived of their wives.
>
> (Isaios 3 [*Pyrrhos*]: 64; Loeb trans., modified)

In other words, the *epikleros'* father's next-of-kin could force the dissolution of her marriage in order to claim her and 'her property' himself – though it has been argued, not very conclusively, that this could not be done if she had already produced a son (Harrison 1968: 309ff.).

Two further points are worth mentioning. First, according to Plutarch's *Life of Solon* (20. 2–3), Solon legislated that the husband of an *epikleros* was required to have sexual intercourse with her three times a month, and that if her husband was incapable of this, then she had the right to 'consort' with her husband's next-of-kin – though Plutarch finds the latter law 'strange and ridiculous'. Second, according to the speaker of Isaios 10:

> though Aristomenes or Apollodoros [respectively the patrilateral uncle and cousin of the speaker's mother, whom the speaker claims to have been a defrauded heiress] could have had my mother adjudicated to them, nevertheless they had no right to her property. Seeing that neither Apollodoros nor Aristomenes, if either of them had married my mother, could possibly have had control of the property – in accordance with the law which does not allow anyone to have control of the property of an *epikleros* except her sons, who obtain possession of it on reaching the second year after puberty.
>
> (Isaios 10 [*Aristarchos*]: 12; Loeb trans., modified)

In other words, the next-of-kin within the *anchisteia* who claimed the *epikleros* in marriage along with her father's property did not have unconditional rights to the property. He merely held it in trust until such time as the son(s) she had borne him came of age, at which point they automatically inherited what was their maternal grandfather's estate.

If these last two points are considered together and in conjunction with the accommodation already noted between the bilateral and patrilineal aspects of the Athenian kinship system, I think something of the rationale of the *epikleros* can be understood. Prior to Solon's legislation, which allowed a childless man to adopt a son and leave him his property, such a man's *oikos* would cease to exist for want of an heir and his property would be absorbed by his closest collateral relative. If he had a daughter only, an *epikleros*, then that daughter would simply go the same way as the property. At most her attachment to the property would ensure that she was not left bereft.

But if Solon's legislation allowing the inheritance of a man's estate by an adopted son was designed to ensure that the number of the

Athenian *oikoi* remained constant (thus, as it were, frustrating the rights of collateral kin), perhaps the role of the *epikleros* can be seen in the same light. Her function is now to supply an heir to her father's *oikos*, and any son she bears becomes the automatic inheritor of his maternal grandfather's estate. Her father's next-of-kin, who claims the *epikleros* in marriage, becomes merely the caretaker of the *epikleros*' father's property until such time as she supplies by her marriage with him a male heir for her father's *oikos*. Hence the regulations which demand that the husband of an *epikleros* should have sexual intercourse with her at least three times a month, or, if he was incapable of this, that she should be allowed to 'consort' with her husband's next-of-kin (who would also, of course, be related to her father). Solon's concern was not to protect a girl from being married to some ancient uncle (as has sometimes been suggested), but to ensure that the collateral relative who claimed her in marriage and enjoyed the use of her father's property was capable of performing his duty towards her deceased father's *oikos* by allowing her to become pregnant and to produce a son to inherit and maintain her father's *oikos*. Indeed, with the *epikleros* we have perhaps the limiting case of 'patrilineal' succession for a man's *oikos* being achieved by what was in fact matrilineal succession. The *epikleros* stands in, as it were, for her non-existent brother until she has produced a son capable of carrying on her father's *oikos*. Her father's next-of-kin provides the seed and receives the benefit of control of the property until such time as this end is achieved.[13]

VII

I have avoided many of the more technical problems arising from the study of Athenian succession and inheritance law. I have, however, taken the time to present its general features, for they have considerable bearing on the position of women within the Athenian *polis*. As has been seen in preceding chapters, women were not independent actors in the world of secular affairs. Their legal status made it impossible for them to be the influential owners or administrators of property, and even in the domestic sphere, their allocated domain, they were scarcely the controllers of their own destiny. Nevertheless, given the nature of the Athenian kinship and inheritance systems, they played, whether they wished it or not, a vital role in the life of the *polis*.

Women were the means by which men could draw themselves together into the same kinship groupings, and the means by which not only wealth, but also religious rights and duties were transmitted. Considerable fortunes could be vested in their name: within the *anchisteai* they could, nominally, inherit all of the deceased's relatives property; they could take with them into their husband's *oikoi* very large dowries; and, we should note, they could even be adopted by a childless man to become daughters – necessarily, then, *epikleroi*. Thus in Isaios 11 the speaker says:

> Stratokles, however, happened to receive an addition of more than two and a half talents to his fortune, for Theophon, his wife's brother, adopted at his death one of his daughters and left her his property consisting of land at Eleusis worth two talents, sixty sheep, one hundred goats, furniture, a fine horse which he rode when he was cavalry commander, and all the rest of his goods.
>
> (Isaios 11 [*Hagnias*]: 41; Loeb trans., modified)

Most importantly, of course, a woman's status – her Athenian parentage on the one hand, her placement within an *oikos* by her relatives on the other hand – was determinant on the production of legitimate children to be citizens of the state and to be the heirs of their father's *oikoi*. And inasmuch as the Athenian bilateral kinship system was manipulated to conform to the desires of patrilineal succession, women were not only the producers of legitimate sons and heirs for their husbands but, in the marriage exchanges between *oikoi*, they could be counted on to provide an alternative supply of heirs for their own natal *oikoi*.

In these ways women played a role integral to the economic transferences and kinship solidarity of the *polis*; but their role was nevertheless entirely passive and always subordinate to the interests of men. They were the means by which transactions and alliances could be effected, but effected between and for the benefit of men. As an example, it may be instructive to explore something of the situation revealed in Isaios 11, from which the quotation was taken.

The childless Theophon has left the vast sum of two and a half talents not to an adopted son, but to an adopted daughter, one of the daughters of Stratokles. But the reason we hear of this adoption and inheritance is not because the daughter of Stratokles was thereby made a wealthy young woman, but because the speaker, a certain

Theopompos, was trying to prove to the court how wealthy the *oikos* of Stratokles, her father, was. It was of course Stratokles, not his daughter, who enjoyed the benefit of the bequest, for Stratokles remained his infant daughter's *kyrios*. That is the point of the passage, and Theopompos goes on to explain that:

> Having had complete control of this property for nine whole years, he (Stratokles) left a fortune of five talents three thousand drachmai, including his patrimony, but excluding the fortune left to his daughter by Theophon.
>
> (Isaios 11 [*Hagnias*]: 42; Loeb trans., modified)

Stratokles was wealthy enough without his daughter's inheritance; but his daughter's inheritance was certainly something he profited from. After all, it was he, not his daughter, who was to be seen riding a fine horse as cavalry commander.

Unfortunately next to nothing is known about Theophon except that he was wealthy, childless, and adopted the daughter of Stratokles to whom he left his fortune. Perhaps he had been fond of the little girl (but she could not have been more than four or five years old at the time of his death). Yet something of the reason for this bequest can still be determined: Theophon was also Stratokles' brother-in-law, the brother of Stratokles' wife. If Theophon had no relatives, or no close relatives, other than his sister, then it was only to be expected that he should benefit the *oikos* of his sister's husband by leaving it his fortune. Indeed, it is possible to speculate that Stratokles had made a calculated marriage by taking the sister of a rich and childless man. At all events, it was via his marriage that Stratokles had become Theophon's *oikeios*, his relative, and thus likely to benefit from his wealth, and it was in the name of his daughter, Theophon's niece, that Stratokles received the benefit of Theophon's wealth. Throughout, it was by women that Stratokles profited; but whatever their role, it was still Stratokles, the man, who benefited. Stratokles took a wife from Theophon (Theophon's sister) and gave him back a daughter (the little girl whom Theophon adopted), but the exchange consolidated Stratokles' economic position.

It might, of course, be argued that neither Stratokles nor his *oikos* could have hoped to benefit permanently from the bequest, since it had been left to a daughter, and since this daughter would eventually pass, along with her adopted father's property, into the *kyrieia* and *oikos* of whomsoever was to become her husband. Under the circumstances, however, it was the best which could be arranged, and not so bad at that. The reason why Theophon could not adopt Stratokles' son, thus permanently benefiting his brother-in-law's *oikos*, is clear: any adopted son had to renounce all rights of inheritance within his natal *oikos*, and Stratokles himself needed an heir for his own *oikos* – besides which, he was himself an independently wealthy man. To have transferred his son from his own *oikos* by adoption to the *oikos* of his brother-in-law, Theophon, would have been to gain Theophon's wealth for his son but to have deprived him simultaneously of his patrimony. Having no other son to transfer, a daughter had to suffice. But at the very least, by remaining the *kyrios* of his daughter, Stratokles could hope to enjoy the control of her property until she was married – and he did so for nine years. Indeed, Theopompos, the speaker, clearly implies that Stratokles increased his own fortune by his administration of his daughter's property, or rather, the property of Theophon to which she had become *epikleros*.

And when it came to marriage, who would claim this *epikleros*? As it happens, we do not know who married her. On Stratokles' death she fell under the *kyrieia* of Stratokles' brother, Theopompos, and Theopompos, the speaker, was by all accounts a consummate rogue. In the speech from which I have quoted, he is actually defending himself against the charge of defrauding his nephew, Stratokles' son, who had also fallen into his charge until he came of age; and he was further accused of having failed to provide dowries for the four daughters of Stratokles, one of whom must have been the adopted daughter of Theophon. We lose sight of all the machinations of this family. But it is possible that if Theophon had no relatives, then even as Theophon's adopted daughter and *epikleros*, Stratokles' daughter could have continued to benefit her father's *oikos* by being given in marriage to one of his relatives – but this is guessing.[14]

It would be wrong to argue that women were considered in Athens only as channels for the transmission of property, or only as the means by which alliances were formed between men. As will be seen in the following chapters, they were much more than that. Nevertheless, among the propertied classes women were continually implicated in a complex game of economic consolidation and dynastic manoeuvring in which they were mere pawns, and this must have affected the way men thought of them. One could argue that their very importance within this game as the nominal owners of property and transmitters of wealth can be seen to have had an adverse effect on their importance as individuals. Had they been theoretically as well as practically excluded from owning wealth, and had they been incapable of facilitating its transmission, then their freedom from the constraints and the designs of men might have been greater. As it was, they were pressed into service.

Admittedly there are some suggestions that rights over *epikleroi*, for example, were occasionally allowed to lapse out of consideration for the feelings of a girl. In Isaios 10 the speaker claims that:

> My father received a dowry when he was engaged to my mother and lived with her; but while these men were enjoying the estate he had no means of obtaining its restitution, for when, at my mother's insistence, he raised the question, they threatened that they themselves would obtain the adjudication of her hand and marry her if he were not satisfied to keep her with only a dowry. Now my father would have allowed them to enjoy an estate of even double the value so as not to be deprived of her.
> (Isaios 10 [*Aristarchos*]:19; Loeb trans., modified)

The speaker's claim is that his mother was a defrauded heiress; thus, ironically, his father, who had taken her in marriage by *engue* with only a dowry, could not press her claims to her paternal estate without risking losing her altogether should her claims prove successful. For as an *epikleros* she could then be demanded in marriage by her father's closest relative. And we are given to understand that for reasons of sentiment her husband was quite unwilling to risk this. It might be noted, of course, that he would also have lost her dowry. But it is still personal attachment to a wife which is stressed, and even if the argument is dismissed as a piece of pleading designed simply to sway the jury in the speaker's favour, it is still significant that this sort of emotional argument was considered valid. The court was expected to understand a man's affection for the woman he had taken in marriage. On the other hand, this speech also illustrates the position in which *epikleroi* were placed; they could be claimed in marriage for purely financial reasons and thereby detached from husbands to whom they were already married. Marriage might not exclude affection, but it was not based on it. There were weightier considerations, as old Philokleon in Aristophanes' *Wasps* points out:

> And what if a dying father bequeaths his daughter and *epikleros* to someone. We don't give a damn for the will all solemnly covered and sealed. We give her to whoever entreats and persuades us.
>
> (Aristophanes, *Wasps*: 583–6)

The girl is at the mercy of her father, her relatives, and the law courts, all of whom will try to decide with whom she shall live.

Whatever the sentiments and affections which Athenian men genuinely might have felt towards their wives and womenfolk, women's entanglement in the serious business of securing and conveying wealth could not help but cast them in the role of mechanics whose own feelings had largely, if not completely, to be ignored. Perhaps many men were in no position to do otherwise. An interesting little tale appears in Demosthenes 57:

> Protomachos was a poor man, but by becoming entitled to inherit a large estate by marrying an *epikleros*, and wishing to give my mother in marriage, he persuaded my father, Thoukritos, an acquaintance of his, to take her in marriage, and my father received my mother in marriage at the hands of her brother, Timokrates of Melite.
>
> (Demosthenes 57 [*Euboulides*]: 41; Loeb trans., modified)

It is difficult to penetrate beneath the surface of so bald an account; but we see a man who was obviously concerned for his wife's future, at least concerned enough to arrange personally for her remarriage to one of his friends. Yet he could still dispense with her, and being a poor man felt he had to dispense with her, in order to be able to claim a large inheritance – an inheritance which would, of course, be accompanied by a new young bride. We are not so far from the situation of Euripides' *Medea*, but it does not appear to have been so remarkable a situation.

I shall quote one last passage. Given the factual complications and *lacunae* of Isaios 6 [*Philoktemon*] it may or may not have been the case that the sister of Philoktemon was an *epikleros*; equally, it may or may not have been the case that the speaker was misrepresenting points of law.[15] Nevertheless, something of the Athenian woman's practical situation when she found herself embroiled in rival claims to the succession of an *oikos* and the inheritance of its property can be judged from its tenor:

> You have, therefore, gentlemen, to consider whether this woman's son ought to be heir to Philoktemon's property and go to the family tombs to offer libations and sacrifices, or whether it should be my client, Philoktemon's sister's son, whom he himself adopted; and whether Philoktemon's sister, formerly the wife of Chaireas, and now a widow, ought to pass into the power of our opponents and be married to anyone they choose, or else be allowed to grow old in widowhood, or whether, as a legitimate daughter, she ought to be subject to your [the court's] decision as to whom she ought to marry. These are the points you have now to decide by your verdict.
> (Isaios 6 [*Philoktemon*]: 51; Loeb trans., modified)

The dispute is about property, and it is a dispute between men. But the implications for at least one woman, the sister of Philoktemon and the widow of Chaireas, are considerable, for depending on how the facts of the matter were interpreted by the court, and depending on how the property was eventually settled, then either this woman could find herself under the guardianship of her nephew (the speaker's client), or be transferred to the guardianship of the speaker's opponents to be married off to whomsoever they pleased (which the speaker implies would not be a satisfactory arrangement for her) or, worse, be neglected by them and allowed to grow old in widowhood, or else she could be found to be an *epikleros* and thus subject to the court's decision as to who was her closest male relative with the right to claim her in marriage. The speaker is adducing these alternative fates for the sister of Philoktemon in an effort to sway the court. But they are nevertheless incidental considerations – minor consequences in a case not being fought to determine the future of a widow, but one being fought to determine which men should inherit an estate.

6

FREEDOM AND SECLUSION

I

The laws of the Athenian *polis*, framed and instituted by men and officially defining the role of women, are important indices both of women's actual position within Athenian society and of the manner in which the female was perceived. Laws are *par excellence* the means by which society attempts to give itself explicit form and to regulate the conduct of its members in accordance with accepted morality (cf. Gould 1980: 43). But laws are also rigid and incomplete reductions of that morality and embrace only a limited field of activity. The axioms of right and wrong are usually left unstated. Many of the constraints imposed by society find no mention in official ordinance, while what might formally appear to restrict the actions and influence of a particular group can often in practice be bypassed, modified, or forgotten, if only because the exercise of the law, more concerned with specific proscription than general permission, comes into play only at those crisis points when individuals are forced to appeal to the officially stated prohibitions of society rather than to the tacitly accepted allowances of daily life (cf. Post 1940: 421–2).

In law Athenian women were most certainly subordinate to men. They were consigned to the background of events to be protected, controlled, and manipulated by those who held the monopoly of authority in a society which was, by definition, a society of men. But how accurately does such a picture reflect the true state of affairs? I have already attempted to supply something slightly more than a reconstruction of certain parts of the law *per se* by presenting it in the actual process of its operation where its effect on the lives of women can be registered and its enunciation in the mouths of men can be heard. (The bulk of legal evidence comes in any case from law-court speeches rather than from codified statutes – which, for the most part, have not survived.) But even if an account of the legal procedures involving women does accurately illustrate a genuine part of Athenian life (and not so many dead letters from unexercised legislation) it could still be argued that it illustrates only those exceptional moments when the law had need to operate, leaving unrecorded the more normal times when it did not. It remains to be seen, then, whether women, whose position in law made them at all times dependent on men as their masters, protectors, and representatives did at all times lead lives in conformity with this legal subordination. If the law refused to see women as independent beings, did they lack independence in their daily lives? And if the law saw women as always sheltered within the *oikos* of their *kyrios*, did this legal seclusion translate itself into domestic confinement?

II

> There was no question of a girl being free to meet other young people, since she scarcely ever left the women's apartments, the *gynaikon*. Whereas married women seldom crossed the thresholds of their own front doors, adolescent girls were lucky if they were allowed as far as the inner courtyard, since they had to stay where they could not be seen – well away, even, from the male members of their own family.

Thus writes E. R. Flacelière (1965: 55), and his portrayal of women's seclusion is by no means unrepresentative of much scholarly opinion. It is, however, fairly clear that Flacelière's picture is based largely on Xenophon's treatise on household management, the *Oikonomikos*, and on certain salient passages from the orators – evidence which, by contrast, the optimists, Gomme and his followers, have tended to gloss over. In fact I think such evidence is of extreme importance and I shall return to it. Nor do I believe that Flacelière's account is entirely mistaken. But it must be admitted that there is a significant body of evidence which will not square with a picture of rigorous physical confinement.

First, there are the numerous women mentioned in Aristophanes' plays: petty traders and retailers in the market. Admittedly, as Ehrenberg noted (1974a: 150), metics and foreigners were greatly involved in Athens' trade, and it is thus difficult to know how many of these women were non-Athenians. The woman who ran the inn/brothel where Herakles got himself into so much trouble in *Frogs* was certainly a metic,[1] but the indications are that others were the wives and daughters of citizens, for when Lysistrata organizes her women's strike to end the war, she cries:

> Forth to the fray my warrior women allies.
> O egg-and-seed-and-potherb-market-girls,
> O garlic-selling-barmaid-baking-girls.
> (Aristophanes, *Lysistrata*: 456–8; Loeb trans., modified)

Lysistrata's strike is a strike to persuade the Athenian *polis* to come to terms with Sparta. It could therefore be assumed that these market women are wives of Athenian citizens. Further, Euripides, the tragedian, was insultingly referred to as the son of a greengrocer woman,[2] and in the *Thesmophoriazusai* (446ff.) it is without doubt a citizen woman, an *aste*, who talks at length about having had to support her family after her husband's death by weaving and selling garlands. Indeed, Aristophanes' plays present a scene in which the Athenian *agora* was thronged with ribbon-sellers, bread-sellers, vegetable-sellers, etc. Whatever their status, they were certainly female; but there is also good reason to believe that many were Athenian.[3]

Nor, in Aristophanes' plays, do the wives of Athenian citizens seem to find much difficulty in crossing the thresholds of their own front doors. In the *Lysistrata* (327ff.) the chorus of women complains about the problems of filling their pitchers with water 'at the spring by the side of the hill', because of the early morning crowd of slaves and other rogues. And Richter is surely right in commenting that in the *Ekklesiazusai* (310ff.):

> Blepyros, discovering Praxagora gone in the pre-dawn hours, is enraged that he cannot find his clothes, suspects the worst reason for her absence, but is far more concerned about being seen in his wife's cloak and slippers than surprised by her leaving the house, even at that hour.
>
> (Richter 1971: 6)

Evidence other than comedy also supports the view that women were not rigorously confined to the house. Somewhat surprisingly Herfst's early study, *Le Travail de la Femme dans la Grèce Ancienne* (1922), seems to show that female labour was not extensively used in agriculture (or rather, it shows that Athenian sources made little mention of it). The peasant farmer's wife appears to have occupied herself with domestic rather than agricultural tasks. Nevertheless, those tasks took her regularly outside the house. In Euripides' play, Elektra, who has been married off to a peasant, complains of her domestic chores – and they include drawing water from the stream,[4] and in Menander's much later comedy, the *Dyskolos*, young Sostratos encounters the girl with whom he falls in love when she too is engaged in this same task.[5] Moreover, in what appears to be the only mention of women being hired for agricultural labour (Demosthenes 57 [*Euboulides*]: 45), it is nevertheless stated that *many citizen women* have been forced to become grape-pickers – an occupation which certainly exposed them to full view.

Perhaps more revealing of the ordinary state of affairs are the speaker's words in Demosthenes 55 [*Kallikles*]:

> Before they undertook this malicious action against me, my mother and theirs were intimate friends and used to visit each other, as was natural since they both lived in the country and were neighbours, and since, furthermore, their husbands had been friends when they were alive. Well, my mother went to see theirs, and the latter told her weeping what had happened and showed her the results; this, men of the jury, is the way in which I learned all the facts.
>
> (Demosthenes 55 [*Kallikles*]: 23–4; Loeb trans., modified)

Certainly in this case the speaker's mother was not confined to the house. As he says, the mutual comings and goings of his mother and her friend were a normal part of rural life. Nor, in this respect, does life appear to have been greatly different in the city. In another law-court speech (Lysias 1 [*Eratostenes*]), Euphiletos' wife (who is deceiving him with a lover) slips out in the middle of the night. When her husband enquires the next morning why the outside door was banging, she simply tells him that the baby's lamp had gone out and that she had gone next-door to relight it. This is presented as a plausible excuse and a natural occurrence (cf. Richter 1971: 6).

There is also clear evidence that women attended certain state ceremonies and speeches. Perikles, after all, did address them in his famous funeral oration. And, at least according to Plutarch:

> When Perikles returned home after subduing Samos he had funeral honours paid to all those Athenians who had lost their lives in the campaign, and he won special praise for the speech he delivered over their tombs, according to the usual custom. As he stepped down from the rostrum, many of the women of Athens clasped his hand and crowned him with garlands and fillets like a victorious athlete. Elpinike, however, came up to him and said: 'This was a noble action, Perikles, and you deserve all these garlands for it. You have thrown away the lives of these brave citizens of ours, not in a war against the Persians or the Phoenicians, such as my brother Kimon fought, but in destroying a Greek city which is one of our allies.' Perikles listened to her words unmoved, so it is said, and only smiled and quoted to her Archilochos' verse: 'Why lavish perfumes on a head that's grey?'
>
> (Plutarch, *Perikles*: 28; Scott-Kilvert trans. 1960: 194, modified)

Not only are Athenian women present to idolize the hero of the hour, but there is also one, Elpinike, Kimon's sister, who was willing to stand out and take Perikles to task on an issue which was both moral and political. Nor, according to Plutarch, was this the first time that Elpinike had had words with Perikles, for she was supposed to have been instrumental in persuading him to propose a decree recalling her brother from exile and, prior to that, to have interceded with Perikles on Kimon's behalf when he was about to be tried for treason.[6] That time Perikles' response had been, 'Elpinike, you are too old, much too old, for this kind of business.' In fact, Plutarch not only has the women of Athens crowning Perikles with garlands; he also repeats the apparently contemporary jibe that Pheidias, the chief architect of Perikles' building programme:

> arranged intrigues for Perikles with free-born Athenian women when they came (to the Akropolis) on the pretext of looking at the works of art.
>
> (Plutarch, *Perikles*: 13. 9–10)

Even if this is a piece of comic slander, it is still an indication that Athens' free-born women were not locked away.

The related question of whether or not Athenian women attended the theatre is a vexed one. On intuitive grounds one cannot help feeling that many of the comedies would have been infinitely more funny if the audience had been mixed, and while Haigh in his classic work *The Attic Theatre* reiterates the conventional belief that 'undoubtedly Athenian women were kept in a state of almost Oriental seclusion' (1907: 324), he also comes close to proving that they were allowed to attend the theatre.

Three passages from Plato all reflect the assumption that women could be spectators of all kinds of drama (Dover 1972: 16).[7] Plato goes so far as to say that the more refined sort of women preferred tragedy. And apart from the oblique indications that women attended the theatre, there is the late and no doubt apocryphal tale that during the performance of Aischylos' *Eumenides* the appearance of the Furies was so horrific that some women had miscarriages.[8] On balance it seems that at the theatre, too, and whether at the performance of comedies or tragedies, women were to be found in the open.[9]

Further, as everybody admits, women attended religious festivals and ceremonies. The *Thesmophoria* and the *Haloa* were exclusively female celebrations and sexual segregation was of their essence; but to take the example of Athens' greatest religious ceremony, celebrated every four years, the *Panathenaia*, women joined the procession as priestesses and as the bearers of ritual objects and were almost certainly present as spectators. For a girl to be chosen as *kanephoros*, 'basket-bearer,' in the procession was the greatest honour that could be bestowed on her.[10] Women were also initiated into the mystery cults about which we know so little, but where they surely mingled with male initiates.[11]

Finally, weddings and funerals provided two frequent occasions at which women were prominent. The care of the deceased, the washing and anointing of the corpse, was very much a woman's task, indeed her prerogative. In Isaios 8 the speaker states that:

> I presented myself, accompanied by one of my relatives, a male cousin of my father's, to convey away the body with the intention of conducting the funeral from my own house. . . . When, however, my grandfather's widow requested that the funeral should take place from that house [i.e. the deceased's] and declared that she herself would like to help us lay out and deck the corpse, and entreated me and wept, I acceded to her

request and went to my opponent and told him in the presence of witnesses that I would conduct the funeral from the house of the deceased since Diokles' sister had begged me to do so.
(Isaios 8 [*Kiron*]: 21–2; Loeb trans., modified)

And, as pictorial evidence clearly shows, during the formal lamentation and the funeral procession to the cemetery, women took the major role as mourners of the dead (Alexiou 1974: 5ff.). Indeed Solonian legislation, which will be discussed later, was forced to limit, though by no means to proscribe, women's participation in the public lamentation of the dead for fear of the very disturbance it created.

As for weddings, the principal rite was a bridal bath for which a procession of women went to fetch water from the spring, the *Kallirhoe*. Again the scene is illustrated by vase paintings: a crowd of women holding torches, and among them a flute player, march ahead of one woman who bears the *loutrophoros*, the container in which the water is to be brought back. The bride was anointed with unguents by her female attendants, and at the sacrifice and feast given by the bride's *kyrios* the bride was attended by her female friends and seated beside the *nympheutria*, a woman whose task it was to guide and help her throughout the marriage ceremony. She was then conveyed to the groom's house by waggon or ox-cart followed by her relatives and friends, male and female, singing the marriage song and with the bride's mother carrying the carriage torch (Flacelière 1965: 62ff.). For women, marriages probably constituted one of the major recurrent social events in their lives – and certainly one in which they were out in public.

III

Much of the evidence to which I have referred above has been adduced by the 'optimists' to counter the thesis of female seclusion presented, for example, by Flacelière (1965). Richter actually argues that, 'The goings on we read about in the comedies and court speeches would surely lead the sedate moralist to wish for greater surveillance. The young wives were as undisciplined a bevy of nymphs as Hellas ever reared' (Richter 1971: 7). Certainly the evidence makes it very difficult to subscribe to the idea that in Athens women were quite literally locked away, never or seldom to be seen outside their house. It seems clear that in Athens' daily life, in the market, in the fields, visiting neighbours, at public meetings, at the theatre, at religious festivals, at funerals and at weddings, and even during the performance of their daily domestic duties, women were to be seen. But does this entirely rule out the idea of seclusion and segregation? In deviating from Flacelière's view, is it necessary to leap to the opposite extreme and imagine Athens' women and girls cavorting in gay abandon, free from all constraints?

The first point to be made is that, even if discussion is confined to *astai*, to citizen women, with foreigners, metics, and *hetairai* left aside, women were not a single group (Pomeroy 1975: 60). The very young and the very old probably enjoyed a greater degree of freedom than those who, to underline the significant feature, were possessed of their full sexuality. It is stated, for example, that a commonplace tactic of Athenian law-court pleading was for the accused to present his wife, children, and friends to the jury in an effort to arouse the jury's pity. Yet, in the references to this practice that are generally cited, there are only two examples where those placed before the jury include women: the famous passage from Aristophanes' *Wasps* (568ff.) in which Philokleon talks of a father displaying before the jury his little boy and his little girl, and the passage from Demosthenes 25 [*Aristogeiton*] in which the speaker says:

> The sight of the children of some of the defendants and their aged mothers standing in court does not move him to pity.
>
> (Demosthenes 25 [*Aristogeiton*]: 84; Loeb trans., modified)

Evidence may simply be lacking for other sorts of women, but I think it more than probable that the women whom defendants were willing to make a spectacle of before the courts tended to be their infant daughters or their aged mothers, not their wives.

Similarly, in the passage from Demosthenes 55 [*Kallikles*] quoted earlier, it is probable that the speaker's widowed mother had rather more freedom to go off visiting her neighbours than any adolescent daughter whom the speaker might have had. In this context it is worth recalling the replies which Perikles is alleged to have made to Elpinike, for in both cases he remarks on her age, and in both cases there is a sexual allusion. Perikles seems to be implying that had Elpinike been younger and more attractive, then she might have proved a better advocate for her brother by turning her sexuality to advantage. Conversely, however, Elpinike's very ability to approach Perikles in so straightforward a manner may well have resulted from the protection age afforded her from sexual slander.

Second, it is reasonable to assume that there were significant differences between the conduct of women from the upper economic classes and those who belonged to the poor. When the speaker of Demosthenes 57 [*Euboulides*] (who sold ribbons in the market-place with his mother) states that many women worked as grape-pickers, he specifically mentions that this was:

> owing to the misfortune of the city in those days; and many who were poor then are now rich.
>
> (Demosthenes 57 [*Euboulides*]: 45; Loeb trans., modified)

This may have been a piece of special pleading. It may also have been the case that in practice the employment of women in agricultural tasks (as well as in nursing and spinning)[12] was not a great oddity. But the speaker is certainly claiming that his mother was caught up in an abnormal state of affairs, and he is certainly subscribing to an ideology, in terms of which his mother's exposure in the market-place was a shameful misfortune in just the same way as other women's employment as grape-pickers was an unfortunate, if necessary, breach of respectable social convention (cf. Schaps 1979: 62). The Athenians' attitude towards employment and wage-labour will be examined in a later chapter, but for the moment it is enough to note that the speaker would very much have preferred his mother not to have been forced into what he saw as a 'slavish' situation, and he feels it incumbent on himself to offer an apology for this state of affairs. Similarly, it should be noted that the references to Euripides' mother having sold vegetables in the market-place, whether true or false, were definitely intended as an insult. What the women of the poor were forced to do need not contradict a dominant ideology in terms of which female seclusion was desirable. And such an ideology can remain dominant, although it was perhaps only the well-to-do who could translate it fully into practice.

This brings me to the third and most important consideration. In dealing with so complex a social phenomenon as sexual segregation and female seclusion, we are indeed dealing with an ideology, with a set of moral ideas and values. These could still operate although women went to the spring to draw water, or slipped out of the house to visit neighbours, or even went on sight-seeing tours of the Akropolis. To try to determine whether or not Athenian women were secluded and segregated by whether or not they appeared outside their houses or talked to men is to adopt an extraordinarily simplistic criterion which would be rejected by anyone who has actually spent time in, for example, modern rural Greece. Interestingly, for the most part it would be an inapplicable way of registering female seclusion and sexual segregation even in those Islamic or 'oriental' societies which have been taken by classical scholars as the archetypal examples of female confinement. Feminine seclusion does not have to be enforced by bolts and bars. It can be ensured by the constraints of a morality in terms of which the worlds of men and women are separate, divided by conceptual as much as physical boundaries, and in which the distance between men and women is measurable in terms of familiarity and 'respect' as much as in terms of physical proximity (cf. Gould 1980: 48–9).[13] With this in mind, let me turn to Xenophon's *Oikonomikos* and to Ischomachos and the education of his wife.

IV

Ischomachos, a wealthy country squire, is interrogated by Xenophon's Sokrates on the subject of 'economy' – of household and property management – because by all reports Ischomachos is the very model of the successful gentleman farmer. Nor does Sokrates find himself disappointed in Ischomachos.

It is important to remember, however, that Xenophon is presenting Ischomachos and his relationship with his wife as an *ideal*. It is clear that Ischomachos does not subscribe to the common view of women, and that he is very much a free-thinking liberal in his treatment of his wife – in the degree of authority he delegates to her and in the affection he shows towards her. I suspect Ischomachos' views would now strike most readers as appallingly condescending. Be that as it may, neither Ischomachos, nor Sokrates, nor presumably Xenophon and his contemporaries, saw Ischomachos' behaviour as anything but admirable, and it is perhaps worth noting that even in 1945 Werner Jaeger could write:

> Our idea of Greek womanhood would lack many of its finest and most essential features if we did not possess Xenophon's picture of the education of the wife of an important country

> squire. . . . [Here we have] the ideal of a woman thinking and acting for herself, and having her own broad sphere of influence, as described by Xenophon in the best traditions of country life and civilization.
>
> (Jaeger 1945: 175–6)

Sokrates encounters Ischomachos in the market-place, the *agora*, and asks him how he passes his time, since, by his appearance, he obviously spends most of it out of doors. Ischomachos concurs:

> Well, now, Sokrates, since you ask the question, I certainly do not pass my time indoors; for, you know, my wife is quite capable of looking after the house herself.
>
> (Xenophon [*Oikonomikos*]: 7.3; Loeb trans., modified)

Sokrates, greatly interested in the subject of Ischomachos' wife, then asks him whether he educated her himself or whether she had received instruction from her mother and father. Ischomachos replies:

> Why, what knowledge could she have had, Sokrates, when I took her? She was not yet fifteen years old when she came to me, and up to that time she had lived under the utmost care, seeing as little as possible, hearing as little as possible, and saying as little as possible. If, when she came to me, she knew no more than how to turn out a cloak when given the wool and had seen only how the spinning had been given out to the maids, is that not as much as could be expected?
>
> (ibid., 7.5)

Sokrates then enquires whether Ischomachos taught his wife the rest of what she now knows, and Ischomachos happily begins his painful rehearsal of the education of his wife:

> Well, Sokrates, as soon as I found her docile and sufficiently domesticated to carry on conversation, I questioned her to this effect:
>
> 'Tell me, wife, have you realized for what reason I took you and your parents gave you to me? For it is obvious to you, I am sure, that we would have had no difficulty in finding someone else to share our beds. But I for myself, and your parents on your behalf, considered who was the best partner for home and children that we could get. My choice fell on you,

> and your parents, it seems, chose me as the best they could find. Now if the gods grant us children we shall then consider how best we shall educate them. For one of the blessings we shall share is the acquisition of the very best of allies and the very best support in old age; but at present we share in this, our *oikos*. For I am paying into the common stock all I have, and you have put in all that you have brought with you.'
>
> (ibid., 7.10–13)

Duly impressed, Ischomachos' wife nevertheless protests that she can see no way in which she could be of practical help to her husband and his *oikos* since her mother had always told her that her sole task was to be *sophron* (discreet/chaste/modest/well-behaved). Everything, therefore, remains up to him. Not so, claims Ischomachos, who explains that the gods in their wisdom coupled man and woman in order that they might form a perfect partnership in mutual service. Human beings, he says, require the shelter, protection, and storage space of the home as well as the produce that comes into it from outside:

> And since both indoor and outdoor tasks require labour and attention, woman's nature was from the start, I think, adapted for indoors and man's for the outdoor tasks and concerns.
>
> (ibid., 7.22)

Ischomachos then goes on to outline the nature of this division of labour which, he asserts, is both the design of the gods and sanctioned by law. And he adds:

> Thus for the woman it is more seemly to stay indoors than to abide in the fields, but for the man it is unseemly to stay indoors rather than to attend to the work outside.
>
> (ibid., 7.30)

Ischomachos then pursues the analogy of the queen-bee and her hive in order to impress on his wife both the importance of her duties and the glory she will gain from them. She is to supervise all domestic affairs, children, and slaves, and to keep everything in order within the house from which he and others will go out, returning with the bounty which she must again guard, manage, and distribute. Thus she will win the esteem of the whole *oikos*. Indeed:

> the pleasantest experience of all is to prove yourself better than I am and to make me your servant; and, so far from having cause to fear that as you grow older you may be less honoured in the household, to feel confident that with advancing years, the better partner you prove to me and the better guardian of the children of the house you become, the greater will be the honour paid to you in the home.
>
> (ibid., 7.42)

The rest of Ischomachos' account concerns a more detailed description of his wife's duties: she must teach spinning and weaving to the slave-girls; she must keep all the household equipment in order and neatly arranged; she must praise or admonish the household slaves and regulate their diet; she must participate in their work herself and not sit round growing idle and fat; in short, she is to be the queen of her hive. And, we are led to understand, Ischomachos' wife delights in her new role.

It should be noted, however, that Ischomachos nowhere says that his wife must not leave her house. The gist of his whole argument is that there are separate but complementary spheres of activity for men and for women, and these constitute precisely those things which are outside (*ta exo*) and those things which are inside (*ta endon*). Ischomachos is referring predominantly to a division of labour and to a division of competence and authority – not to a physical boundary which would imprison his wife within the walls of her house. There is thus no reason to presume that his wife never left her home, that she did not attend festivals, that she was not to visit her neighbours, or even that she could not journey to the city. What is being stated is that her *concerns* relate exclusively to the home, and that the home is a sanctuary from the outside world over which she is to preside.

On the other hand, Ischomachos clearly says that it is seemly for a woman to remain inside, whereas it is unseemly for a man to do so. Here something rather more than a division of labour is touched on. The sphere within which a woman is to be competent also becomes the sphere in which it is at least *appropriate* for her to remain. And a minor point should be noted. When Ischomachos is explaining the details of his wife's duties, he says:

> We know, I take it, that the city as a whole has ten thousand times as much of everything as we have; and yet you may

> order any sort of household slave to buy something in the *agora* and to bring it home, and he will not be at a loss. Everyone of them is bound to know where he should go to get each article.
> (ibid., 8.22)

As it happens Ischomachos is mentioning the *agora* to his wife only because it provides a convenient analogy for the manner in which all household articles should be stored and arranged in their separate places. But the incidental implication is that Ischomachos' wife would not normally go herself to the city; she would send out a slave (and, of course, Ischomachos, being a wealthy man, had plenty of slaves). Moreover, whatever the frequency of her occasional sorties, once outside the protection of the *oikos* Ischomachos' wife is quite literally 'out of place'. She would be entering the man's world, the public world, where she would lack both authority and protection – where, unless she betrayed herself as something other than what she should be, she would have neither the right nor, in Ischomachos' terms, the ability to assert and to govern herself.

Furthermore, both Ischomachos' remarks and those of his wife leave little doubt that as an unmarried girl her life had been a guarded one. She was to see, hear, and say as little as possible. According to her own mother, her task was simply to be *sophron*. The virtues she knew were negative ones. She had been brought up to be what was once called 'innocent' – possessed of a literal unworldliness. Ischomachos changes that, but only by defining for her the scope of a world within which she is to be, and must be, competent. As a matron she will become mistress of her own house; but out of her house her unworldliness is to remain.[14]

V

Perhaps I have slightly exceeded Ischomachos' stated words in my interpretation, but what I have said is consistent with them and allows the evidence of Xenophon's *Oikonomikos* to be integrated with the sort of evidence produced by the 'optimists'. Sexual segregation and female seclusion in Athens did not necessarily entail the rigorous confinement of women within the women's quarters of the house. It did, however, mean the ideological separation of the domestic world of women from the public world of men. And it meant that if women were outside the house, then a boundary ought still to separate them from unknown males – a boundary largely maintained by the mutual awareness of men and women of the *social* distance between them.

Such a form of sexual segregation can be real enough, and the constraints imposed on women's freedom of behaviour still quite rigorous. It is not, however, open to contradiction by the mere fact that women could be seen outside their homes. And although I think it may be safely presumed that the greatest factor in ensuring the observation of 'correct' social conduct between the sexes was men and women's own commitment to the conventions of respectability – that is to say, their own internalized moral values – probably those same conventions also meant that in public upper-class women did remain in groups with their kinswomen and friends, or under the gaze of some close male relative, or accompanied by an attendant slave-girl (Lacey 1968: 168).

It is interesting to note, for example, the case presented in the speech in Hypereides 1 [*Lykophron*]. Lykophron, a middle-aged man of some standing, was accused of having seduced the sister of Dioxippos and of making her pregnant while her first husband was an invalid. After her husband's death she was remarried to a certain Charippos. The prosecution adduced reports of relatives which stated that during her wedding to Charippos, Lykophron had followed the bride and tried to persuade her to have nothing to do with her new husband but to reserve herself for him. At his defence, Lykophron waxes particularly indignant at this accusation, for not only does he claim the story to be baseless; he also claims that it is patently absurd:

> Is there anyone in Athens so uncritical that he could believe these allegations? There must have been attendants, men of the jury, with the carriage that conveyed the bride: first a muleteer and a guide, and then her escort of boys, and also Dioxippos [her brother]. For he was in attendance too, since she was a widow being given away in marriage. Was I then so utterly senseless, do you think, that with all those other people in the procession, as well as Dioxippos and Euphraios his fellow wrestler, both acknowledged to be the strongest men in Greece, I had the impudence to pass such comments on a free woman and was not afraid of being strangled on the spot?

Would anyone have listened to such remarks about his sister as these men accuse me of having made without killing the speaker?

(Hypereides 1 [*Lykophron*]: 5–6; Loeb trans., modified)

The jealousy with which a free woman's reputation would be guarded by her male relatives is made clear. To impugn her honour, and thereby to impugn the honour of the men under whose guardianship she resided, was a deadly offence. We might agree with Lykophron that, all things considered, the story was rather unlikely. But another point should be noted. If, for the reasons Lykophron adduces, the story was false, why did not his accusers concoct a more credible tale? That, for example, he had continued to see his alleged mistress at the market-place or nearby her house and had been overheard making his proposals to her. Part of the reason may have been dramatic effect. What more scandalous than the scene of a lover following and entreating his mistress at the very moment of her marriage? But perhaps there are other considerations which, *pace* Lykophron's protestations, would have made the story not so incredible to Athenian ears: that despite the presence of attendants, the muleteer, the company of boys, and her brother, Lykophron's mistress was more in the open, more readily approachable during her marriage procession, than she had been at any other time since the death of her invalid husband. Dangerous as it was, Lykophron may have been seizing his only chance to have words with his mistress.

In this context it might be remembered that, according to the speaker's account of the seduction of his wife in Lysias 1 [*Eratosthenes*], the alleged seducer, Eratosthenes, did not approach the speaker's wife directly. He accomplished his seduction through the intermediary of a slave-girl whom he approached after a funeral and who subsequently became his messenger. Eratosthenes was in a position to see and admire the speaker's wife, but he was in no position to be seen talking to her. Here, then, it was a woman's public appearance at a funeral (as opposed to a wedding) which aroused an amorous male. As the speaker complains, his mother's death marked the start of all his troubles. Opportunities for familiarity with a free woman thus appear to have been scant and tales of seduction coincide precisely with those special occasions on which it is clear that women were away from the *oikos* and in public view:

weddings, funerals, religious and state festivals. The speaker of Lysias 1 goes on to relate that while he was away in the country his wife went off to celebrate the women's festival, the *Thesmophoria*, in the company of none other than Eratosthenes' mother. That a man's wife had the freedom to go off during his absence to celebrate these important female rites in the company of an elderly matron appears normal; but that this festival should further a seducer's designs also implies that other occasions on which she would be away from the house and legitimately in the company of people other than her kin were rare. It is perhaps worth noting that it is such relatively unguarded moments that Menander appears to have fixed on in order to further the contrivance of his plots (Gould 1980: 47): in the *Epitrepontes*, during the festival of the *Tauropolia*, with its nocturnal revels, Charisios rapes a girl (who later turns out to have been his own wife); in the *Samia*, during the roof-top festivities of the *Adonia*, Moskhion seduces a neighbour's daughter (whom he then desires to marry but whose child creates some extraordinary domestic confusion).[15]

The degree of supervision and protection under which Athenian women remained was not, therefore, inconsiderable. But they were closely watched rather than locked away. And this, as Gould (1980: 40) has shown, is the situation portrayed in tragedy as well. There is no gross contradiction between the conventions of social respectability to which the women of Athens appear to have been subject and the purported 'freedom' of the heroines of tragedy. To borrow two of Gould's examples in Euripides' play, Elektra's peasant husband strongly criticizes his noble wife (after the arrival of Orestes and Pylades) in the following words:

> Well, who are these strangers I see by my door? What brings them to my rural door? Are they looking for me? It's shameful for a woman to be standing with young men.
>
> (Euripides [*Elektra*]: 341–4)

In Sophokles' *Elektra* Klytaimnestra similarly rebukes her daughter 'for wandering untethered' in public view and 'bringing shame on her friends' by rejecting Aigisthos' male control. If Elektra's conduct does not amount to an escape from prison, it is still a serious breach of social convention. She has failed to conform to the retiring role which befitted women in their dealings with the outside world and unrelated males. Nor does Klytaimnestra herself entirely evade the rules of accepted behaviour. When in Euripides' *Iphigenia in Aulis*, she approaches Achilles to talk of the marriage of her daughter, Achilles' first reaction is one of acute embarrassment at this open encounter with an unknown and free-born woman (830ff.). In Aischylos' *Agamemnon*, Klytaimnestra addresses the chorus of Argive elders, and though her dominance has already been turned into a quite explicit psychological peculiarity by the reference made to her 'male mind', the chorus still comments that when the king is away and the male throne vacant, then it is right to honour the woman – as if by way of an 'historical' gloss to explain to the Athenian audience the very possibility of the scene.

Moreover, a morality which required women to avoid familiarity with men and which posited the home as woman's proper place clearly meant that women were at least supposed to spend the better part of their time indoors. In Aristophanes' *Ekklesiazusai* the women of Athens, disguised as men, take over the *Ekklesia* and vote to transfer control of the *polis* to women. The real men who attend the *ekklesia* that day are afterwards somewhat perplexed by the pallid countenances they have observed.[16] The reaction suggests that the wives of citizens did, as Ischomachos implies, remain largely inside. One of the early conventions of Greek vase-painting had been that of the 'sunburnt man' and the 'shadow-bred' woman, a convention that persisted into the fourth century, to the extent that artists appear to have been unwilling to apply the new techniques of tonal shading to the flesh of women who, in contrast to men, remain blankly white (Robertson 1975: 425). But caution is necessary here, for this is a matter of convention, both of artistic representation and of human beauty (and there are few subjects more convention-ridden than the latter). It cannot be argued securely that the pallid complexions of the women either in Aristophanes' plays or on vase-paintings are proof that women remained indoors. Women whitened their faces with cosmetics, and they carried parasols to protect themselves from the sun[17] (Lacey 1968: 168; Pomeroy 1975: 83). The real point is that the conventions of Athenian female beauty related to an *ideal* of female seclusion. White-faced women and suntanned men were symbolic representations of male and female roles. Such symbolism is of extreme importance in underlining what was held to be appropriate, but not necessarily in revealing what was at all times the case. It is a statement in aesthetic terms of a morality to be found in practice, but not necessarily limiting practice. Paint and parasols could always patch things up.

Nevertheless, the ideal of female seclusion and the notion that a man's *oikos* and the women who resided in it constituted an inviolable domain protected from the outside world were strongly felt. They are clearly evident in two passages addressed to the male juries of a law-court. In the first, from Lysias 3 [*Simon*], the speaker had come to blows with a certain Simon, and Simon charged the speaker with wounding with intent to kill. The speaker, in defending himself, relates the cause of their affray and attempts to blacken the character of his accuser:

> I think it proper that you should hear the numerous offences that he has committed against myself. Hearing that the boy was at my house, he came there at night in a drunken state, broke down the doors, and entered the women's rooms. Within were my sister and my nieces who had lived such well-ordered lives that they were embarrassed to be seen even by their relatives. This man, then, carried insolence to such a pitch that he refused to go away until the people who appeared on the spot and those who had accompanied him, thinking it an awful thing to do to intrude on young virgins, carried him off by force.
>
> (Lysias 3 [*Simon*]: 6–7; Loeb trans., modified)

But does this passage also mean that the speaker's sister and nieces were permanently locked up in the women's quarters? I think not. The speaker is impressing on the jury the outrage that Simon has committed. In this context his narration of Simon's drunken invasion and appearance before the girls in their room tellingly symbolizes the violent profanation of his private world; and it would call into play all the protective sentiments of those whom he was addressing. It would arouse all the moral indignation of every right-minded male juror who knew that it was his duty to maintain and protect the nice and the innocent – his wife, his daughters, his sisters, his nieces. And the speaker leaves no doubt that his womenfolk were such modest and retiring creatures – that they shrank from the gaze even of those who were their relatives.

Gould (1980: 47) provides a brilliant commentary on the second passage, from Demosthenes 47 [*Euergos*], a complicated case concerning the equipment of a trireme. The *Boule* passed a resolution requiring all trierarchs to retrieve missing gear from their predecessors, and the speaker of Demosthenes 47 thus arrived at the house of Theopompos to demand restoration of what was missing. Theopompos was not at home, and the trierarch waited outside while a slave-girl was sent to fetch him. When Theopompos returned, he insulted the trierarch and, since the ship's gear was not forthcoming, the trierarch eventually entered Theopompos' house to seize securities. As Gould points out, at this stage in his narrative to the court the trierarch pauses to explain (1) that the door was open (he was not breaking in), and (2) that he knew that Theopompos was not married. 'It is the brief aside we should notice: even in this crisis a self-respecting Athenian is not going to run the risk of coming face to face with another man's wife in his own home' (1980: 47). In a subsequent and contrasting episode, the shameless Theopompos and his brother-in-law break into the trierarch's farmhouse and confront his wife, children, and an old nurse who are eating in the courtyard. The remaining female slaves lock themselves in the tower. But when violence breaks out and the shouting attracts the attention of neighbours' slaves and of a passing neighbour, Hagnophilos, no one enters the house. Indeed, Hagnophilos did not go in because, as the trierarch explains, 'he did not think he was justified in the absence of the *kyrios* of the family from the house'. As Gould concludes:

> Few passages, perhaps, bring out so clearly the sense of an inviolable boundary separating the free women of a household from unrelated males and of the outrage implicit in male entry upon the women of another kinship group.
>
> (1980: 47)

In sum, then, Athenian women were not literally locked away. Nevertheless, the ideology of female seclusion, the degree of male oversight, and the workings of a morality of social distance all meant that women's legally subordinate and sheltered role was matched by the mores of daily life. In the tale of Hypereides 1 [*Lykophron*], in which Lykophron is defending himself against a charge of seduction, the legal and the social subordination of women can be seen nicely intertwined. As already noted, the speech illustrates the degree of jealous oversight to which a woman's chastity and honour could be subjected by male relatives: 'Would anyone have listened to such remarks about his sister as these men accuse me of having made without killing the speaker?' says Lykophron. It is a man's duty to guard, protect, and defend his womenfolk's virtue. But if what appears to have been the motivation for Lykophron's indictment is also considered, then all is not a matter of high moral sentiment. Lykophron's alleged mistress, the sister of Dioxippos, had born a son, and that son, who was about three years old at the time of the trial, stood to inherit the entire *oikos* and property of his late father, the first husband of the sister of Dioxippos, to the complete exclusion of collateral relatives. Until that son came of age, the *oikos* of the first husband of the sister of Dioxippos was in the charge of a certain Euphemos, who had been appointed guardian by will. Now it may be that at first this will was accepted, or it may be that some relatives disputed it from the start, or it may be that the relatives hoped that the little boy would not survive infancy and that, as members of the *anchisteia*, they would then inherit the property. At all events, by the time of the trial it seems that all the relatives united in an attempt to gain the property of the first husband of the sister of Dioxippos, and the only way to do that was to prove that the little boy was not his father's son, that he was not legitimate, that he was a *nothos* and could not inherit. Hence the indictment of Lykophron for adultery, an indictment that tried to prove that his alleged affair with the sister of Dioxippos had preceded her first husband's death, that, indeed, Lykophron had tunnelled through the walls of her husband's house to have intercourse with her. In short, the only way to gain the inheritance was to destroy Lykophron's reputation and charge him with adultery – and, of course to destroy the reputation of the sister of Dioxippos (a woman who had already, we might note, been married off by her brother to a man of his choosing after her first husband's death). Thus not only was such a woman subject to all the constraints that a morality of female seclusion, chastity and good-behaviour required. She was also subject to all the machinations that an inheritance system involving women and based on the principle of legitimate succession afforded those whose interests were material.

7

PERSONAL RELATIONSHIPS

I

Any attempt to penetrate the *oikos* itself and to assess the quality of personal relationships within it, the degree of intimacy and affection existing between its members, and thus the amount of influence or power which a woman might have had over husband, father, brothers, or sons runs the risk of exceeding the limits of available evidence and the area within which worthwhile sociological generalization can be made. Little is known about life within the *oikos* and, on *a priori* grounds, it would be foolish to expect that unequivocal statements could be made about what are, by definition, individual relationships. Nevertheless, snippets of information exist that reveal something of the *public* ideals of private life – and, on occasion, they also afford glimpses of its reality.

As noted earlier, sentimental appeals to the welfare of women (and children) formed part of the stock in trade of Athenian oratory (Lacey 1968: 175). Such appeals were used to various effect. Apollodoros has already been heard attempting to stir the (male) jury in his prosecution of Stephanos and Neaira by drawing attention to the plight in which free-born Athenian women would find themselves were Stephanos and Neaira to be acquitted – an attempt which also relies on the threat of the shame which the male jurors will feel in the face of their wives' and daughters' indignation. But at the beginning of this same speech, Apollodoros' son-in-law, Theomnestos, also makes repeated allusion to the interests of women – to those of his own female relatives:

We have suffered grievous wrongs at the hands of Stephanos,

and have been brought by him into the most extreme peril – my father-in-law, myself, my sister and my wife.

(Demosthenes 59 [*Neaira*]: 1; Loeb trans., modified)

[People] flung in my teeth the charge that I was the most cowardly of men, if, being so closely related to them, I did not take vengeance for the injuries done my sister, my father-in-law, my sister's children, and my own wife.

(ibid., 12)

Similarly, in Demosthenes 48, the speaker complains:

I am striving not only in my own interest, but in the interest of her to whom I am married, his (the opponent's) own sister born of the same mother and the same father, and in the interest of his niece, my daughter. For they are being wronged not less than I, but more. Can anyone say they are not suffering outrageous treatment when they see this fellow's *hetaira*, in defiance of all decency, decked in masses of jewels . . . while they themselves are too poor to enjoy such things? Are they not suffering a wrong even greater than my own?

(Demosthenes 48 [*Olympiodoros*]: 54–5; Loeb trans., modified)

In Demosthenes 23 the speaker glosses Drakon's homicide law (see pp. 53–4) with the following:

He allows the man who slays anyone so treating (raping or seducing) any of these women (wife, mother, sister, daughter, *pallake*) to go completely free, and that acquittal, men of Athens, is the most righteous of all. Why? Because in defence of those for whose sake we fight our enemies to save them from indignity, he permitted us to slay even our friends if they insult and defile these women in defiance of the law.

(Demosthenes 23 [*Aristokrates*]: 55–6; Loeb trans., modified)

In the first passage above, the speaker, Theomnestos, is aiding Apollodoros, to whom he is doubly related, being both his brother-in-law and his son-in-law. Apollodoros had married Theomnestos' sister, Theomnestos Apollodoros' daughter, his own niece. His kinship connections therefore obliged him to come to Apollodoros' assistance, for their households were joined and they were *oikeioi*. This is clearly stated by Theomnestos. But what Theomnestos stresses in this passage is not simply his obligation to enter into a prosecution in conjunction with his father-in-law, thereby upholding the honour of their related households, but his obligation to assert the honour specifically of his sister (Apollodoros' wife), his sister's children (Apollodoros' children), and his own wife (Apollodoros' daughter). The women who create the links between Theomnestos' *oikos* and Apollodoros' *oikos* are included as interested parties *in their own right*, and the court is asked to take their feelings and standing into account as well those of the men in whose charge they reside. As for the speaker of Demosthenes 48, what he wishes the court to consider is not only his own poverty in the face of the immoral extravagance of his brother-in-law, but also the hardships, material and psychological, that the situation he describes has created for his wife and daughter. Finally, in the last passage the speaker goes so far as to suggest to the male jury (as if it were a matter of common knowledge) that the reason why wars are fought by men is to protect the dignity of women.

In three differing types of speeches – a prosecution motivated by personal vengeance and concerning Athens' endogamic marriage laws, a law-suit in pursuance of an inheritance, and a political prosecution – references to the welfare of women come readily to the speakers' lips. No matter how self-interested their real motives, and no matter how rhetorical the references to women, it remains true that Athenian speakers employed such emotive appeals and doubtless did so on the assumption that they would prove effective.

Rhetoric of this sort reveals much about the Athenians' protective attitude towards women which, far exceeding mere jealous possessiveness, could assume a sentimental concern for their rights and happiness (as determined, of course, by men). It does not, however, reveal a great deal about individual relationships in practice. Rather more important in this respect are the sort of statements made, for example, in Isaios 7 and Demosthenes 50:

> so he came to my mother, his own sister, for whom he had a greater regard than for anyone else, and expressed a wish to adopt me, and asked her permission, which was granted.
>
> (Isaios 7 [*Apollodoros*]: 15; Loeb trans., modified)

> My mother was Apollodoros' sister, and a close affection never interrupted by any quarrel existed between them.
>
> (ibid., 43)

> While I was abroad my mother lay sick and at the point of death. She had often sent for me before, begging me to come by myself if I could not come with my ship. My wife, too, to whom I am deeply attached, was in poor health for a long time during my absence.
>
> When I heard these facts from the lips of those who arrived, and also through letters from relatives, how do you think I must have felt and how many tears must I have shed while I reckoned up my present troubles and was longing to see my children and my wife and my mother, whom I had little hope of finding alive? For what is sweeter to a man than these, or why should one wish to live if deprived of them.
>
> (Demosthenes 50 [*Polykles*]: 60–3; Loeb trans., modified)

These passages once again come from public law-court proceedings, and the speakers are reciting sentimental tales to gain far from sentimental ends. Nor is there any means to check their true feelings, or the feelings of those to whom they refer. Nevertheless, it is important to note that despite the staunchly male nature of Athenian democracy speakers had little hesitation in declaring before an all-male court the degree of their devotion towards women. And even if, in context, such statements amounted to no more than calculated appeals to male sentimentality, it remains true that Athens was a society in which such a form of sentimentality was collectively endorsed and could therefore be publicly played on.

But in the above examples something more than an attitude of paternalistic protection is also revealed. The speakers are talking of their own and other men's emotional *dependence* on women. Women are not merely the recipients of care and affection. In Apollodoros' words women (and, importantly, children) are what make life worth living. It is not so strange, then, that the speaker of Isaios 7 can present to the court as evidence of the validity of his adoption by his uncle (his mother's brother) the fact that his mother and her brother had been on close and affectionate terms and that the adoption had taken place with her full blessing. Despite the fact that *legally* his mother would have had no say in the matter, the speaker is clearly assuming that an all-male court will both recognize the importance of a woman's views in family matters and, in arriving at its decision, take into account what he represents as her desires.

Such expressions of affection for women, such admissions of dependence on women, and such public recognition of women's influence are all far removed from contempt. If women were consigned to the background of events and to the private world of each citizen's *oikos*, this is not to say that the male Athenian necessarily considered the world of his *oikos* and his women to have been irrelevant to his own happiness and emotional fulfilment. Even when, as in comedy, the familiar note of misogyny is sounded, in many cases it serves more to betray the closeness of familial relationships and the practical influence of women than to express a simple disregard for the female sex. One such example occurs in Aristophanes' *Wasps*. Philokleon returns home from the male world of public affairs with the pay he has received for jury service secreted in his mouth (a common form of monetary transport):

> O then what a welcome I get for its sake;
> my daughter is foremost of all,
> and she washes my feet and anoints them with care
> and above them she stoops and lets a kiss fall
> till at last by means of her pretty 'Papas'
> she angles my three obols pay,
> then my little wife sets out on the board
> breads in a tempting array
> and cosily seats herself close by my side
> and entreats me to try this and that.
> (Aristophanes, *Wasps* 606–11; Loeb trans., modified)

The passage is, of course, quite waspish and it is comedy. Philokleon is raising a laugh, not praising his daughter and wife. But if the passage did tickle an Athenian audience, then presumably they must have recognized in it the parody of a familiar scene: a scene which mixes intimacy, affection, and female connivance in equal measure.

In a more serious context, in Lysias 32 a widow's children had been defrauded by their guardian, her late husband's brother (who was also her own father, i.e. she had been married to her paternal uncle). This she did not suffer lightly, and

> entreated me [the speaker] to assemble her father and friends

> together, saying that even though she had not before been accustomed to speak in the presence of men, the severity of their [her children's] misfortunes would compel her to give a full account of their hardships.
>
> (Lysias 32 [*Diogeiton*]: 11; Loeb trans., modified)

According to the speaker's account that follows, the widow then gave a very creditable performance, upbraiding her father in no uncertain terms before the assembled audience, and demonstrating a remarkably detailed knowledge of the family's financial affairs (Lacey 1968: 160–1). But it should be noted that the speaker, no doubt anxious to create the right impression in the minds of the male jurors, carefully underlines the fact that, despite her initiative and defiance of her father's authority, the widow was not, of course, 'accustomed to speak in the presence of men'. It is difficult not to see this as simply a public gesture.

Other examples could be adduced, but all confirm in a fragmentary way that in individual cases the relationship between a man and his wife, sister, mother, or daughter could be extremely close and could also allow a good deal of *de facto* female authority.[1] In comedy the latter possibility becomes quite explicit, since almost by definition comedy involves a breaking or reversal of the rules and norms of accepted conduct. Weak husbands are 'funny' because they are supposed to be strong; clever slaves amusing because they are supposed to be stupid. But to be found funny rather than simply absurd, such reversals and transgressions must be recognized to contain an element of truth. When, for example, in Aristophanes' *Clouds*, a simple farmer, Strepsiades, is dominated by the aristocratic lady whom he has married and is financially ruined by the son who takes after her, then comedy is surely admitting some unofficial realities.

II

Pictorial evidence must also be deemed relevant for an understanding of the place that women held in the affections of men; but equally, it is extraordinarily difficult to interpret, for it is all too easy to read into a mute work of art those feelings one considers appropriate and then to attribute them to the figures presented. Nevertheless, sculptural representations confirm the closeness of the bonds which existed either actually or ideally between men and their womenfolk. The best illustrations are the carved reliefs of private tombstones that commemorated women or that included women in the scenes portrayed. In one respect their evidence is unambiguous, for however erroneous the interpretation of an outstretched hand or an inclined head might be as the outward sign of some genuine emotion, it is nevertheless the case that the mere existence of such monuments attests the importance of women in the private lives of men and proves that in Athens it was considered fitting to acknowledge that importance in a public way.

From the third quarter of the fifth century BC down to the late fourth century there is an unbroken series of sculptured funeral reliefs. They depict women almost as frequently as men.[2] One of the best early examples is a monument to a certain Ampharete who is shown holding 'a bird in the hand on the chairback, and on her knee with the other hand a baby, whose round head and raised hand issue from a shapeless bundle of clothes. She looks down at it with what seems an air of sadness' (Robertson 1975: 366). Inscribed on the monument is the verse: 'This is my daughter's dear child I hold, whom when we lived and saw the sun's rays with our eyes I held on my knee, now too I hold when we are both dead.' It is a moving inscription and a moving work of art, and though the effect of both must be credited to the poet and the sculptor, it must still have been commissioned (and paid for) by a male relative, presumably by Ampharete's husband, or son, or perhaps (given the inscription) her son-in-law. And it should be noted how much it is a tribute to women. Despite the strongly patrilineal character of the *oikos*, the dead child whom Ampharete holds is referred to only as 'my *daughter*'s child'. Grandmother, daughter, daughter's child – this is the line sadly celebrated by the art of a male society.

Such reliefs depicting women only – for example, that of Hagnostrate, daughter of Theodotos, from the late fourth century, or that of Hegeso, daughter of Proxenos, with her slave-girl, or another which shows one woman seated with her face raised and reaching out to touch another woman who bends to stroke her cheek (with a little slave-girl in the background)[3] – are important indications of the esteem in which women were held. Their deaths (or lives) are as publicly and expensively marked as those of men. Taken on its own this pictorial evidence would incline one to think that women were in every way the social equals of men. And it is important to note that these monuments are not representating female deities or women engaged in public religious activities (such as in the Parthenon), where rather different reasons could be adduced for their appearance. They are privately erected monuments celebrating ordinary, if wealthy, women, and it must be assumed that the motivation for their construction lay in some combination of genuine affection and regard and the persuasions of a collective social morality that deemed it appropriate, if not obligatory, that women should find their share of posthumous honour.

More suggestive, however, are the reliefs depicting both men and women. Domestic scenes were common on these monuments. As Robertson implies (1975: 379), Athenian funeral reliefs seem as much connected with life as with death, and where men and women are shown together there are glimpses of what were, or what were held to be, relationships within the *oikos*. Indeed, in many of the multi-figure reliefs it is difficult to know which of the people portrayed had actually died (Humphreys 1983: 106). In one sense these representations do little to contradict the evidence of other sources, literary or legal. Women are shown in their prescribed domestic role, as mothers of children, as brides of men. At times a somewhat patriarchal note is struck, as with one bearded figure who sits 'like a god enthroned' (Robertson 1975: 380), apparently presiding over the women who stand around[4] – though this is by no means always the case, as the reliefs that depict only women sufficiently demonstrate. On the other hand, although marriages in Athens were both contracted and dissolved on grounds which appear to have had little to do with the personal sentiments of individuals, the pictorial evidence seems to suggest that the bonds that united husband and wife could be both enduring, and, if we give ourselves the slightest licence for interpretation, loving. The relief of the couple Thraseas and Euandria with their hands clasped surely indicates as much, even if, in death, what is depicted is the idealization of a relationship rather than its actuality.[5] Similarly, the family group reliefs that contain husband, wife, daughters, and sons illustrate the closeness and unity of the *oikos* in terms of what one would be inclined to see as manifest mutual affection.

Robertson's comments on the lack of distinction between the living and the dead in these reliefs, and on the common motif of the clasped hands, are of considerable interest. The motif of the clasped hands

> has been interpreted as a gesture of farewell, or of reunion; there are examples which either fits well, but neither fits all. Either or both may sometimes have been present to the mind of artist and client, most often perhaps the farewell, suggested by the overtone of sorrow in attitude and expression; but the basic idea is probably a more general one of union. We saw the same motive used on record reliefs to symbolize the union of states. Here it is the unity of the family, across death; and that accounts too for the lack of a clearly defined distinction between living and dead. Possibly such family-stelai were set up, with one name inscribed, when one member died, and as others were laid in their long home other names were added. Death is a line in time which sooner or later all will cross, and these representations are in some degree timeless.
>
> (Robertson 1975: 379)

The timelessness of the *oikos* – or rather a man's desire to maintain his *oikos* in perpetuity through a line of direct legitimate male descendants, natural or adopted – was embedded in the law, morality, and religious belief of the Athenians. But here a facet of this same ethos can be detected, which is not so apparent in the law or the other areas of Athenian social practice recoverable from texts. The continuity and survival of the *oikos* is no longer expressed solely in terms of an enduring male establishment into which women were merely contracted as temporary members for the purpose of procreating legitimate male heirs. Rather, women can now be seen depicted as themselves, integral to the *oikos*, bound into it by ties of mutual affection and esteem which are to survive beyond life itself.

There is perhaps little surprise in this, and if such glimpses of family life seem to contradict what has been discussed of women's position in the earlier chapters, then that may be only because there is still a tendency to assume that limitations imposed on the freedom of women can be equated in some simple way with a disrespect or contempt for women. The opposite can be just as true, and no matter how much we would now want to reject the attitudes and practices of Athenian society, the very constraints imposed upon women, the scrutiny to which their behaviour was subjected, and indeed their exclusion from the public world and exile within the protected domain of the *oikos*, could all be construed, and most probably were construed by Athenian men, as indications of a concern for their welfare rather than as uncaring dismissal. Nevertheless, despite the respect in which women publicly appear to have been held (at least while contained within their domestic role), despite the genuine affection which seems in many cases to have existed between a man and his dependent womenfolk, and despite the real if illegitimate power that women could sometimes exercise over their menfolk (all of which can be illustrated by oratory or literature or art), the definition of a woman's place as being within the house while a man's lay outside it could still result in a distanced relationship between husband and wife – or at least, it could supply all the preconditions for estrangement.

III

Xenophon's account of Ischomachos and his wife (see pp. 114–18) is preceded by a conversation between Sokrates and a certain Kritoboulos. Sokrates asks:

> Is there anyone to whom you commit more affairs of importance than to your wife?

The reply is 'No', and the conversation proceeds:

> Is there anyone to whom you talk less?
> Few or none, I confess.
> And you married her when she was a mere child and had seen and heard almost nothing?
> (Xenophon, *Oikonomikos*: 3.12–13; Loeb trans., modified)

Perhaps too much has sometimes been made of this passage; it may well be that Kritoboulos' relationship with his wife was an extreme. Nevertheless, it does serve to underline an important aspect of male/female relationships in Athens, and one which has already been touched on in general terms at the outset of this book: the barrier created between men and women by the very variety of extra-domestic male concerns within the democracy. Although Sokrates refers to the internal organization and running of the *oikos* as 'affairs of importance', Kritoboulos' external pursuits leave him little time to spare for his conjugal life. The consequent delegation to women of authority and competence in domestic affairs, a division of labour that has already been mentioned, has resulted in this case in personal estrangement. Kritoboulos scarcely speaks to his wife. He follows his own interests and leaves her to get on with her proper concerns. Sokrates, we might note, does not approve of this lack of communication, which is precisely why he goes on to recount the exemplary relationship of Ischomachos and his wife. If the affairs of the household are of importance, then a man should concern himself with them and with the 'education' of his wife. But for many men, and especially for those of the upper classes, the range of social and political interests outside the home could easily have reduced 'wives' to exactly what Apollodoros defined them as: bearers of legitimate children and guardians of *ta endon*, 'the things inside'. They cared for a man's *oikos*, but they were not his chosen companions in life, for life was not passed within the home.

Nor, importantly, were the extra-domestic interests of Athenian men confined to politics and matters of public concern. Sex, too, could be an extra-domestic pursuit. With Sokrates and his circle of friends, where all male social life is encountered in its best attested form, philosophical matters are found mixed with a discreet homosexuality. But further, it should be noted that in the passage quoted (on p. 123) from Lysias 3, in which the speaker relates to the court the outrageous entry of Simon into his house and Simon's unseemly encounter with its women, the quarrel between Simon and the speaker (which gave rise to the disturbance) was in fact over a Plataian *boy*. A speaker might wax indignant before the court about the invasion of his domestic world and the terrible affront thereby given the sheltered women in his charge, but it is fairly clear that the speaker's own erotic interests lay outside his home and not necessarily with women. This, in somewhat evasive language, is also conveyed to the jurors, but such embarrassment as the speaker expresses does not relate to homosexuality.[6]

To take an example of a rather different sort, let me return to Lysias 1 where the speaker is trying to convey to the court a picture of his entirely satisfactory marital life prior to his wife's seduction by Eratosthenes. As he explains:

> When, Athenians, I decided to marry and brought a wife into my house, for some time I was disposed neither to vex her nor to leave her too free to do just as she pleased. I kept watch on

> her as far as possible and with such observation as was reasonable. But when a child was born to me, thenceforth I began to trust her and placed all my affairs in her hands, presuming that we were now in perfect intimacy.
>
> (Lysias 1 [*Eratosthenes*]: 6; Loeb trans., modified)

What the speaker meant by 'vexing' (*lypein*) is not entirely clear: perhaps no more than that he did not overly supervise her and place too many domestic duties on her. But there is also the possibility of a reference to sexual demands (Pomeroy 1975: 83). At all events, the 'intimacy' to which he refers is presented as being the direct result of his wife's having given birth. The closeness of the relationship and the trust he places in her are not bound up with any (stated) feeling of growing affection towards her as a person. Rather, their 'intimacy', as the Greek term *oikeiotes'* implies, has to do with the joint establishment of a domestic unit and with the procreation of children. The speaker has taken a wife, and she has fulfilled her role by running his home and bearing a son. Their unity is expressed in these terms. It is perhaps also worth noting that after the speaker's wife had given birth, she was said often to have slept on her own with the baby in a downstairs room so as not to disturb her husband. And though it may be unfair to claim that this was anything other than what it was presented as being – a practical arrangement – neither, in the light of other evidence, is it difficult to see it as an example of the fact that a wife's sexual role was primarily procreative rather than erotic (cf. Pomeroy 1975: 83).

The degree to which most Athenian men did derive their emotional and sexual satisfaction from extra-marital relationships, whether heterosexual or homosexual, is difficult to gauge (and I shall return to the question of homosexuality in due course). But as Dover has pointed out, the plot of Aristophanes' *Lysistrata* could hardly have been conceived had not most Athenian men relied for the most part on their own wives for sexual satisfaction (1974: 211).[7] On the other hand, if only to show the manner in which comedy can stress or suppress certain facets of life to achieve its comic purposes, in Aristophanes' *Ekklesiazusai* one of the reforms proposed by the women who have taken over the *polis* is to rid the city of prostitutes so that they, Athenian wives all, might enjoy the monopoly of their husbands' favours. Nor is there any doubt that prostitutes and brothels abounded in Athens. As Xenophon's Sokrates remarks to young Lamprokles:

> Of course you don't suppose that sex (*ta aphrodisia*) provokes men to beget children when the streets and stews are full of means to satisfy that. Obviously we select as wives the women who will bear us the best children, and then marry them to raise a family.
>
> (Xenophon, *Memorabilia*: 2.2.4; Loeb trans., modified)

Ischomachos' comments to his new young wife (pp. 115–16) imply much the same thing, and whether or not most Athenians frequently patronized brothels, or slept with slave-girls, or kept *hetairai*, it seems clear that the purpose of marriage, in social as well as strictly legal terms, was to beget legitimate children, not to supply men with exclusive sexual partners. As Dover summarizes:

> No danger attended the sexual use of women of servile or foreign status, whether they were prostitutes owned by a brothel-keeper, *hetairai* who were looking for long-term dependence on agreeable and well-to-do men, concubines owned by the user himself or lent by a relative or friend, or dancers, singers, or musicians whose presence at men's drinking parties exposed them to importuning, mauling, kidnapping (an occasion for fighting between rival males), temporary hire, or straightforward seduction enjoyed by both partners.
>
> (1974: 210)

The observations of Xenophon's Sokrates above would justify such an interpretation, but in fact the evidence for all manner of sexual licence outside the home with women of non-citizen status is overwhelming.

The illustrations of painted pottery, though no doubt in many cases intended to have an explicitly 'pornographic' character, nevertheless depict (among other sexual scenes) the erotic moments of male drinking parties which were well recognized social institutions and for which slave-girls, foreign courtesans, and *hetairai* were hired as sexual entertainers. Indeed Xenophon's *Symposion*, (dominated though it is by a discussion of 'love' which is both homosexual and moralistic) closes with a young slave-boy and girl creating an erotic tableau of the scene of Psyche's seduction by Eros. But it is fair to add that the effect of this performance on the young men who were watching was to send those who were married hastening back to their wives and to persuade those who were not to take wives forthwith (while Sokrates, for his part, went for a walk).

In comedy the same sexual exploitation of non-citizen women is frequently portrayed. The provision of brothels for travellers is enthusiastically referred to in Aristophanes' *Frogs* (113), and Dionysos even imagines himself in the role of a slave looking on and masturbating while his master has his way with a prostitute (542–8). At the close of the *Thesmophoriazusai* Euripides effects the escape of his friend by appearing with a 'flute-girl' who performs naked, and whom Euripides allows the Skythian 'policeman' to take inside for copulation. Contrived though the whole scene may be, it at least bears witness to the availability of such women and to the casualness with which they could be used. Again, in the *Acharnians* (729ff.), the Megarian pigseller attempts to sell his two daughters to our Athenian rustic, Dikaiopolis – though, to be sure, the daughters are disguised as piglets and the humour of the scene relies on the ambiguity of exactly what the Megarian is offering for sale in a scene replete with extended puns on the word '*choiros*', meaning both 'pig' and 'vulva' (Dover 1972: 63–5). Moreover, it is possible that in the above scenes of the Skythian and the flute-girl and the Megarian and his piglets (and in other such similar 'comic' episodes), actual slave-girls were exhibited on the stage for the delectation of a male audience (cf. Dover 1972: 27).

In just such a scene in Aristophanes' *Wasps* the old reprobate Philokleon brings the play to a close by appearing drunk with a 'flute-girl', while being angrily pursued by the other guests of the party from which he had abducted her – much to the annoyance of his sober son, Bdelykleon. Apart from the verbal relish with which the flute-girl's physical attractions are described, much of the humour of the finale derives from the breaking of social conventions: the kidnapping of an entertainer for purely private enjoyment; the older man indulging in riotous behaviour – hence the comic reversal of a son (normally granted greater social licence) having to chastise his father (who ought to be setting an example of responsible sobriety). But even if the humour does derive from such role-transgressions, the play again shows that sexual entertainers such as the flute-girl were an institutionalized part of extra-domestic male gatherings, and that a fair degree of tolerance was afforded at least to younger men.

The availability of prostitutes for hire, the total acceptance of female non-citizen prostitution, and the licence given younger males to engage in sexual adventures (which at times led to fights and brawls) are also revealed in the more serious evidence of law-court oratory. In one prosecution Demosthenes can state that the defence is bound to take the line that:

> there are many in the city, sons of respectable men, who in sport, in the manner of young men, have given themselves nicknames . . . and that some of them are in love with *hetairai*; that his own son is one of them and has often given and received blows over an *hetaira*; and that's the way young men behave.
>
> (Demosthenes 54 [*Konon*]: 14; Loeb trans., modified)

Similarly the speaker of Lysias 3, charged with assaulting Simon in what was clearly homosexual rivalry, can nevertheless defend himself by drawing an analogy with what he presents as commonplace heterosexual disputes:

> It would be intolerable if in all cases of wounds received through drunkenness or horseplay or abuse or in a fight over an *hetaira* . . . you are to inflict a punishment of such awful severity.
>
> (Lysias 3 [*Simon*]: 43; Loeb trans., modified)

And as Dover comments, ' . . . the mauling and pulling of a slave-girl, with the imminent intervention of someone who wants to take her to a different destination, is not an infrequent motif in late archaic and early classical vase-painting' (1978: 57).

In Apollodoros' lurid account of Neaira's career as a *hetaira*, there is a very frank description of the extra-domestic sexual activities of the relatively well-to-do male:

> When he [a certain Phrynion] came back here [to Athens from Corinth] bringing her [Neaira] with him, he treated her without decency or restraint, taking her everywhere with him to dinners where there was drinking and making her a partner in his revels; and he had intercourse with her openly whenever and wherever he wished, making his privilege a display to onlookers. He took her to many houses to parties and among them to that of Chabrias of Axione, when, in the archonship of Sokratidas he was victor at the Pythian games with the four

> horse chariot which he had bought from the son of Mitys, the Argive; and returning from Delphi he gave a feast at Kolias to celebrate his victory, and in that place many men had intercourse with her when she was drunk while Phrynion was asleep, among them even Chabrias' slaves.
>
> (Demosthenes 59 [*Neaira*]: 33; Loeb trans., modified)

Apollodoros, of course, appears to be highly critical of these goings on. His tone is one of moral outrage. But here the context is important for, as has been seen in earlier chapters, Apollodoros is prosecuting the Athenian citizen Stephanos and the Corinthian courtesan Neaira for living together as man and wife *contrary to the law*. Apollodoros' indignation is directed not so much towards the actual libidinous activities of Neaira and her friends, as against the fact that such a woman is now claiming to be an Athenian wife and the mother of legitimate children of Athenian citizen status. Hence the telling final remark that even Chabrias' *slaves* had had intercourse with her. Athenian society perfected a quite precise double standard which did not involve contradiction, or subterfuge, but rather relied on a clear separation between two categories of women who were not to be confused: legitimate wives (or potential legitimate wives), and all other women. The first were the chaste mothers and daughters of Athenian citizens; the second were open to free sexual exploitation.

On the whole there was nothing straightforwardly immoral about the sexual use of prostitutes and *hetairai* – though this touches on a complex subject to be taken up in the following chapter. Isokrates (3.40) can say that it is *kakia*, 'wrong', to insult a loyal wife by seeking pleasure outside the home (cf. Dover 1974: 210), but the evidence overwhelmingly suggests that sexual pleasures outside the home were made use of with little or no sense of guilt. The consequences of the *excessive* pursuit of such pleasures were perhaps another matter; but that has more to do with the Athenians' belief in the virtue of 'self-control', which will be discussed in the following chapter. In general, respect for the feelings of wives and daughters seemed to entail no more than a degree of circumspection. Thus, again from Apollodoros' prosecution of Stephanos and Neaira:

> When they got here [Athens], Lysias did not bring them [two *hetairai*, Nikarete and Metaneira] to his own house out of respect for his wife, the daughter of Brachyllos and his own

> niece, and for his mother, who was elderly and who lived in the same house; but he lodged the two, Metaneira and Nikarete, with Philostratos of Kolonos, who was a friend of his and as yet unmarried.
>
> (Demosthenes 59 [*Neaira*]: 22; Loeb trans., modified)

As for any consideration for the prostitutes and *hetairai* themselves, the words of the speaker of Lysias 4 make clear the manner in which they were thought of. The speaker and his adversary were involved in a brawl after having jointly paid for a woman. The speaker now complains of his companion:

> he is not ashamed to call a black eye a wound and to be carried about in a litter pretending to be in a dreadful condition, all because of a prostitute – whom he can keep so far as I'm concerned, on restoring the money to me.
>
> (Lysias 4 [*On a wounding*]: 9; Loeb trans., modified)

It is important to remember that this sentiment was openly expressed in a court of law.

IV

Generalizations are indeed difficult to make when it comes to individual (and sexual) relationships. None of the above should be taken to deny the possibility that wives could directly inhibit the promiscuous tendencies of their husbands, and the Athenians certainly did recognize the personal authority that many women held over men. In Aristophanes' *Peace* (1138–9), for example, the chorus of farmers sings of the joys of 'flirting with Thratta *when the wife is away*'. This we might take as a fairly realistic appreciation of the Athenian situation. The slave-girl was fair game for her owner; nevertheless, her owner might be more prone to exercise his rights in the absence of his wife. Similarly, if the tale of Lysias 1 is again returned to, the speaker relates that when he told his wife to go to quieten the baby, she replied:

> 'Yes – so that you can have a go here at the little maid. You pulled her about before once when you were drunk.'

And the speaker continues:

> At that I laughed, while she got up, went out of the room, and

closed the door, pretending to make fun. And she turned the key in the lock.

(Lysias 1 [*Eratosthenes*]: 12–13; Loeb trans., modified)

Here a conjugal scene is presented which is entirely recognizable in its intimate but inoffensive banter of sexual accusation (cf. Gould 1980: 50) – recognizable, that is, except for the fact that a little slave-girl *did* exist who could always be pressed into sexual service by her master.

But even if many an Athenian wife could claim a degree of sexual fidelity from her husband, or at least enjoy a relationship with him whose intimacy could survive the existence of his other sexual liaisons, the effects of separate social spheres for men and women still remain. It is, for example, one of the better known conventions of Athenian social life that a man's wife, sisters, or daughters did not attend even when he entertained his friends at home. A woman's presence at a male gathering could be taken as *prima facie* evidence that she was not a respectable Athenian woman, but a prostitute or *hetaira* (Gould 1980: 48; Dover 1974: 98). Thus, in his prosecution of Neaira, Apollodoros can simply state that:

the defendant Neaira drank and dined with them in the presence of many men as any courtesan would do.

(Demosthenes 59 [*Neaira*]: 24; Loeb trans., modified)

Given Apollodoros' following account (see pp. 140–1) of what took place at the sort of drinking parties which Neaira attended, the conclusion that she was thereby revealed as a prostitute is not surprising (cf. Lacey 1968: 159). Nevertheless, even in less riotous circumstances it appears to have been a firmly held principle of Athenian social life that the presence of a woman when men ate and drank together would mark her as being something less than a respectable wife. Thus, in the case of Lysias 1, the speaker also relates that he happened to meet a friend one day and invited him home to eat; but they ate together in an upper room and not with the speaker's wife. This, it should be noted, was not some raucous drinking party. It was merely a casual meal resulting from a chance encounter. But with a male guest in the house, the speaker's wife remained in her chamber. A man's circle of friends seems to have had little or no contact with the women of his household, and friendships between men and friendships between women appear to have constituted separate networks (cf. Gould 1980: 49). Wives, as we have seen, might visit each other because their husbands were friends or neighbours, but a man did not expect to know socially his friends' and neighbours' wives, nor did these wives mix with each other's husbands.

In this respect a marked contrast is apparent in the behaviour of that other category of women: those who were not Athenian wives, sisters, or daughters and for whom the conventions of sexual segregation and protection did not apply. The term '*hetaira*' may well have been adopted as a simple euphemism for *porne* (prostitute), but the word's literal meaning is 'companion'. Given that, even under relatively innocent conditions, Athenian women were largely debarred from participation in the social lives of men, the use of the term 'companion' for those women who could participate is not without significance. At the least it underscores the social barrier which separated wives from the convivial activities of their husbands. *They* could not be 'companions' in these. Convention demanded that that role be fulfilled by women who were excluded *de iure* from the family and kinship structures of the Athenian *polis*. While it is probable that in the long run most *hetairai* lived wretched lives, it also seems likely that such intellectually accomplished women as existed in Athens were to be numbered among this class. Aspasia is the paradigm case. She was neither an Athenian nor was she Perikles' wife, and most probably she ran a brothel. But not only is Perikles' devotion towards her amply attested – she is spoken of highly by both Plato's and Xenophon's Sokrates for her ability as a conversationalist and adviser rather than simply because of her physical attractions – according to Plato's *Menexenos*, she composed Perikles' famous funeral oration (cf. Pomeroy 1975: 88–90).

Aspasia was no doubt an exceptional figure, but her example alone is enough to underline the fact that any woman who was likely to become the intellectual and, in a sense, the social equal of a man would have to be a *hetaira*; for it was in that capacity alone that a woman could have entry into male society. Ironically, the very respectability and protection enjoyed by Athenian wives meant their exclusion not only from the more lurid activities related in Apollodoros' prosecution of Neaira, but also from the intellectual gatherings and discussions that Plato and Xenophon report. Perhaps Kritoboulos did not talk to his wife on the justifiable grounds that she had little to say that was of interest to him; but, like

Sokrates and Perikles, he may well have found time to discuss affairs with those *hetairai* who were well informed as well as for hire.

It is in this context that the importance of homosexual relationships must be considered. Exactly why certain societies should openly admit the possibility of homosexual pleasures while other societies should deny or suppress them cannot easily be answered,[8] but it is clearly the case that (male) homosexuality was a deeply entrenched part of Greek civilization. As Dover (1978) has shown, however, homosexual relationships were still subject to a variety of social and cultural inhibitions. The sexual, whether homosexual or heterosexual, seems always to be hedged around with prohibitions and constraints; yet, insofar as *under certain conditions* sexual relationships are legitimated by any society, then in Athens homosexuality was as legitimate as heterosexuality.

To take a simplistic point of view, it might seem that the mere acceptance of homosexuality in Athens would tend both to weaken conjugal bonds and to undermine the position of women simply by, as it were, doubling a man's number of potential sexual partners or even by creating a male society which was erotically self-sufficient. In fact such a view would be misleading because it presents homosexuality and heterosexuality as equivalent alternatives, which in Athens they were not. The distinction between Athenian homosexual and heterosexual practices must be carefully noted; the distinction does not, however, lessen the effects of institutionalized homosexuality on the status of women.

Only with respect to their strictly erotic aspects do homosexuality and heterosexuality appear to have been equivalents inasmuch as both beautiful boys and beautiful girls were the objects of male desire and in as much as male as well as female prostitutes were to be found in Athens' brothels. Love poetry (from the late archaic period through to Hellenistic times), Attic comedy, the writings of Plato and Xenophon, and numerous passages from law-court proceedings make it clear that in Athens, and throughout Greece, sexual desire was aroused (and satisfied) by both boys and girls.[9] The numerous *kalos* inscriptions on painted pottery collected by Dover (1978: 111ff.), probably commissioned by male lovers for their loved ones, extol the physical attractiveness of both male and female recipients, though there is a preponderance of the homosexual variety. Xenophon's *Symposion* reveals the virtual interchangeability of girls and boys as the objects of sexual (and aesthetic) admiration. Near the beginning, young Autolykos' beauty dumbfounds the older (male) guests on his arrival while, at the close of the work, an erotic heterosexual performance (see p. 138) sends the diners racing back to their wives. This situation is further and rather more explicitly illustrated by Athenian vase painting. For example, in one of the works of the Nikosthenes potter a youth reclines while a woman practises *fellatio* on him; three other youths queue up behind the kneeling woman to enjoy her in turn; but while the foremost of the three youths titillates the woman from behind, the last youth, perhaps impatient at having to queue, attempts to pull the second youth back onto his own erect penis (Dover 1978: 86–7).

But once one goes beyond the question of what physically aroused the Athenian male to consider the social aspects of homosexuality, then important differences emerge – differences which actually present heterosexual relationships in a comparatively unfavourable light.

As argued earlier, though the well-brought-up Athenian girl might not have been literally locked away she was certainly put beyond the bounds of casual dalliance. Those women who were to be the wives and mothers of Athenian citizens were sexually obtainable only on condition that they were granted the *status* of Athenian wives and mothers. Men married late, generally at about thirty years of age, and premarital sexual relationships – indeed any sexual relationship outside of marriage – had to be with some person other than a citizen woman. Not, as has been shown, that available female partners were hard to come by. But what was lacking for the Athenian male prior to marriage (or even after marriage) were women who were his social equals, partners who could enter into a relationship with him which, though based on sex or on sexual attraction, could at the same time supply or create that refractory complex of social and emotional needs, desires, pleasures, and anguishes which for want of a better word we call love and the Athenians, recognizing its ambivalent properties, classified as a sickness (cf. Dover 1974: 210). Perhaps that was not always the case. It seems some young men did 'fall in love' with their *hetairai*, and it seems clear that a courtesan such as Aspasia could command considerable respect and even deference from the men she knew (though not from the population at large). But, in general, and especially in a society so status-conscious as Athens', a man's premarital (and extra-marital) heterosexual relationships had to be with women who, for the most part, came from that despised and indeed legally dehumanized group – slaves – who thus fulfilled no more than a sexual need. And even if a man's *hetaira* was a free-born foreigner and a relatively wealthy and independent woman, she was still outside that exclusive group that alone was accorded full respect. In any case a relationship with such a courtesan was essentially a mercenary one. One paid for what one got (Dover 1974: 210; Pomeroy 1975: 91).

None of the above, of course, fully explains the existence or acceptance of homosexuality. But, granted its existence and acceptance, in the context of the fairly strict segregation obtaining between men and women of citizen status, it was able to take on a particular function, or rather complex of functions. Homosexual relationships were the sole form of sexual relationship which could, outside the bonds of legitimate marriage, be entered into by partners who were both of the same social station. Furthermore, they were the sole form of sexual relationship in which an element of mutual admiration and respect might be considered fundamental, since they were entered into by both partners through a process of courtship and, in the final analysis, were dependent neither on monetary payment (as in the case of prostitutes) nor on the social necessity of procreation (as in the case of legitimate wives). In other words, they were the sole form of sexual relationship which afforded all the preconditions for what we might term 'romantic love'. And it is worth noting that at such stage as romantic *heterosexual* love makes its appearance in the comedies of Hellenistic Athens, then, as Fantham observes, 'their problem is to reconcile romance with morality' (1975: 46). And since in practice morality forbade love affairs with any women except courtesans and prostitutes, 'tortuous, and in some way shocking, plots were necessitated by the romantic ideal of love sealed by marriage, in a society which all but precluded the combination of the two elements in everyday circumstances' (1975: 56).

Needless to say, not all homosexual relationships could be described as 'romantic'. As mentioned, male prostitution existed, and in fact the penalty for male prostitution by a *citizen* was the loss of all political rights (about which more will be said in the following chapter). Nor was what is commonly described as 'Platonic' love necessarily without its carnal elements, including copulation (Dover 1974: 214). Nevertheless, with homosexual relationships between free men what is always stressed (and idealized) in the sources is the nature of a relationship based upon mutual respect, admiration, disinterested magnanimity, and indeed aesthetic appreciation. What is played down (and in Plato's writings denigrated) is the carnal aspect of the relationship (Dover 1978: 153–70). In many respects the situation was not so different from that which has generally prevailed in our own society with regard to heterosexuality. On the one hand sexual or physical attraction is a normal element in the relationship (if not at first supplying its very basis) and is recognized as such. On the other hand, what is publicly extolled is less the physical pleasure deriving from the relationship than the 'purer' emotions of love and affection held to be generated by it – and to such an extent that the latter has sometimes been made deliberately to obscure the former (Dover 1978: 90). A further analogy could be drawn between the present norms of heterosexuality and Athenian homosexuality. A man might boast about his sexual conquest of a youth but, ideally, the youth was supposed to resist his advances and maintain his chastity. Certainly he did not advertise the fact of his submission (Dover 1974: 215). On the other hand, to be the object of a lover's admiration was considered both personally gratifying and publicly honourable. If this seems a complicated charade then we should recognize that it is no more complicated than that which clothes our own heterosexual practices; and to call it a charade is to do it an injustice, for such 'charades' form the substance of many of our most acutely felt concerns. The elements of deception and contradiction appear only when we ruthlessly attempt to reduce means to ends and to deny those social complexities which in practice always constitute the stuff of life. But the point is that in the Athenian case the social, emotional, and moral complexities of erotic relationships were played out for the most part in the context of homosexual relations where the conditions in which they could flourish were best supplied.

The invidiousness of the contrast between homosexual and heterosexual relationships should thus be obvious. Homosexual relationships could be based on affection and mutual esteem between social equals; heterosexual relationships were either with legitimate wives or with prostitutes and *hetairai* – hired slaves or, at best, foreign courtesans performing a sexual service for gain. Relations with female prostitutes and slaves were bound to be seen not only as qualitatively different from, but indeed as markedly inferior to, relationships freely entered into by men of equal rank (cf. Dover 1974: 210). But homosexual relationships provided a contrast not only with sexual relationships between a man and female slaves, prostitutes, and courtesans; they contrasted equally, though in different fashion, with those relationships entered into with Athenian women of citizen status by the process of legitimate marriage, the purpose of which was not companionship but procreation.

Ideally, however, homosexual relationships were between an older man (the *erastes* who was the dominant or active partner) and a younger man or youth (the *eromenos* who was the subordinate or 'passive' partner). As Dover remarks, 'Since the reciprocal desire of partners belonging to the same age category is virtually unknown in Greek homosexuality, the distinction between the bodily activity of the one who has fallen in love and the bodily passivity of the one with whom he has fallen in love is of the highest importance' (1978: 16). But in the literature which both celebrates and moralizes such relationships the age difference which specified the sexual roles of lover and loved one assumes a further significance. The relationship becomes, in some respects, one between mentor and pupil. The older man assumes the role of teacher and moral adviser; the younger man emulates him, learns from his experience, and admires him for his manly qualities which are presented in social and ethical terms. Conversely, the necessity of maintaining the loved one's respect and admiration acts as an incentive for the older man to comport himself in a manner worthy of approval. In short, the exchange between the two parties is that of physical beauty for moral wisdom.

There is no need, of course, to believe the high moralizing of Platonic writings, or to accept that homosexual relationships were no more than the marriages of true male minds with a little aesthetic pleasure thrown in. But given the standard age difference between the partners, and given that the older partner was thus in some sense representative of what the younger partner would become, it can be appreciated that built into the erotic relationship was a commonality of interests, which could extend to embrace the range of Athenian social, and therefore male, concerns. Nor is it necessary to consider Athenian male society to have been erotically self-sufficient to realize that in a society which on so many *other* grounds excluded women from its valued concerns, homosexuality was still likely to achieve a privileged status even with regard to relationships with a legitimate wife. Among the wealthy and educated, wives were neither expected nor required to bring to a sexual relationship social and intellectual companionship since that, after all, was found elsewhere (cf. Dover 1974: 214–15).

It is not known, of course, how Athenian women viewed male homosexuality. Once again the sources are silent. Homosexual relationships were, however, generally confined to the years before marriage and Dover (1978: 171) has suggested that for this reason 'wives will comparatively seldom have had grounds for fearing that their husbands were forming enduring homosexual attachments.' However, as Dover admits, Kritoboulos, who is extravagant in praise of his male loved one in Xenophon's *Symposion* (4.12–16) – the same Kritoboulos who speaks so rarely to his wife (see p. 135) – is himself a newly married man. But even if at marriage and at the comparatively late age of thirty years or so when a man assumed his full political rights and accepted his full social responsibilities he also turned his back on homosexual relationships and became '*un père de famille*', it is difficult to imagine that the conjugal relationship simply substituted itself for previous homosexual ones. One type of sexual relationship may have superseded another, but their qualitative differences must have remained apparent. Legitimate marriage, the raising of a family, marked a new stage in a man's life: a new role to be fulfilled, a new and different set of relationships to be entered into. It might be argued that so it does in most cultures; but a salient difference remains, and one of some importance in understanding the Athenians' view of women. Whereas in our own culture the role that women play in the lives of men may vary through time – from the elusive object of youthful desire to lover, wife, mother of children, companion, etc. (the exactitude of this sequence or its variations are unimportant) – there is nevertheless a continuity throughout these changes: the same person, or at least the same *category* of person, a woman, can assume all these roles.

In the Athenian case, however, it was not only the role of the person which changed, but also that person's gender. Apollodoros' oft quoted statement is that 'we have *hetairai* for pleasure, *pallakai* to care for our daily bodily needs, and *gynaikes* (wives) to bear us legitimate children and to be the faithful guardians of our household'. To be sure, the statement is imprecise and rhetorical and modern commentators have been inclined to argue that the definitions were intended cumulatively and that wives fulfilled the former as well as the latter functions (e.g. Lacey 1968: 113). But I would not so easily dismiss the more apparent meaning of Apollodoros' words, and if the existence of premarital homosexual relationships is added to the picture it can be seen that the Athenian wife, the Athenian woman of citizen status, had but one role which was *exclusively* hers, the procreation of legitimate children. To put the matter in a slightly different way, the appropriation by homosexuality of 'romantic' love – of a relationship freely entered into by both partners out of a sense of mutual esteem and admiration – might leave women who were slaves, prostitutes, or *hetairai* to be no more than despised bodily conveniences. But equally it could leave those women who were of Athenian citizen status (and to whom social respect was due) to be no more than mothers and guardians of the house.

V

Let me now return to the tale of Ischomachos and his wife. The condescending and paternalistic manner in which he instructs her is, to a modern reader, quite striking. Richter (1971: 4) apologizes for this on the grounds that Ischomachos simply cannot talk to his wife as man to woman; he is forced to address her as a child, for while he is presumably in his thirties his wife is not yet fifteen. 'Paternalism' is thus appropriate; Ischomachos' manner has nothing to do with the general Athenian attitude towards women.

Richter, of course, has a point. Ischomachos *was* addressing his wife as adult to child. But it is still indicative of the Athenians' attitude towards women that females appear to have been married straight after puberty while males did not marry until their thirties. The very normality of the age difference allows a conclusion different from Richter's: not merely that Ischomachos was forced to talk to his wife as a child because indeed she was a child, but that for the purpose of marriage (for which Ischomachos has taken her) it was of no consequence that she was (except physically) a child. Despite the concern he shows for her appearance, Ischomachos is not looking for a lover. Those, of whatever variety, he could find elsewhere – and he objects when his wife puts on cosmetics and elevated shoes, for she is thereby blurring the distinction between herself and a courtesan. Nor, despite the concern he lavishes on her 'education', is he looking for someone to be his intellectual and social companion (those he could find, or had found, elsewhere in male company). Rather, he is looking for someone to fulfil competently a series of domestic duties and to become in time the esteemed mother of his legitimate sons. Without malice, with great respect, and with considerable affection, Ischomachos outlines the role of the ideal wife – but it has little to do with the erotic, the intellectual, or the social life of men. It has to do with the furnishing and maintenance of a domestic retreat. But it should still be remembered that in terms of Athenian social structure and Athenian moral expectations Ischomachos' wife held an honoured position. Perhaps when they died a funeral monument could mark their grave, displaying for all to see a picture of Ischomachos and his wife, their hands clasped, their children by their side looking on.[10]

8

THE ATTRIBUTES OF GENDER

I

In the preceding chapters I have tried to describe women's position within the Athenian *polis* in terms both of the law and of the less formal conventions of daily life. I wish now to approach a more refractory topic, namely, the 'nature' of women.

I must stress, however, that I shall not be engaged in the dubious task of trying to reconstruct the Athenian female 'personality'. By 'nature' I mean simply the set of characteristics, real or imaginary, which in the writings of fifth- and fourth-century Athens men commonly attributed to women as natural to their sex. I shall therefore be dealing once again with a facet of Athenian male ideology and not, or not necessarily, with the reality of Athenian women. But my aim will be to show how the characterization of women's 'natural' propensities reflected or was at least congruent with women's role within the social organization of the *polis* and could thus be seen to justify it. In short, I shall be attempting to show how the same moral values exemplified in the formal structures of the law and in the rules of social conduct are also to be found in the psychological constructs which one half of Athenian society (men) made of the other (women).

II

It is not difficult to compose a list of opposing Athenian masculine and feminine traits: men are strong, women are weak; men are brave, women are fearful; men are magnanimous, women are vindictive; men are reserved, women are loquacious; men are rational, women are irrational; men are self-controlled, women are self-indulgent – and so forth. Of course individual writers dissented from or at least qualified these views, but on the whole they represent roughly what may be found expressed throughout the range of Athenian texts, whether literary, philosophical, or legal.[1] Nor should they occasion particular surprise. In any society the concept of masculinity and femininity are defined by mutual opposition and women tend to be portrayed as what men, ideally, are not. The asymmetry of the evaluation is equally unremarkable. If men hold the monopoly of power then it might be expected that they should appropriate the monopoly of virtue. Thus women are not only 'naturally' different; they are also 'naturally' inferior.

On the other hand, such an attitude need not automatically result in the condemnation of women. If women are innately possessed of those characteristics which mark them as inferior, by the same logic the virtues expected of women are to be judged by different standards. As Aristotle states in the *Politics*:

> A man would seem a coward if he had only the courage of a woman: a woman a chatterer if she were no more reticent than a good man.
>
> (Aristotle, *Politics*: 1277b)

Masculine bravery and feminine fear provide a particular example of the relativity of judgement. That women were virtually incapable of displaying courage was an Athenian commonplace[2] – and in a society which was almost continually at war, but in which women were relegated to an exclusively domestic role, there was indeed little opportunity for women to develop or exhibit that fortitude which, for men, was a necessary ideal (cf. Dover 1974: 96). As Ischomachos explains to his young wife in Xenophon's *Oikonomikos* (6.23–5), nature (*physis*) has equipped the bodies of men and women differently. Men are made to endure the hardships of an outdoor life; women are not. According to Ischomachos, however, it is not only in their bodies but in their mind (*psyche*) too that women are differently endowed. If Ischomachos takes it as natural that women should be more fearful than men, it should be noted that throughout the *Oikonomikos* he is also at pains to give women credit for the feminine virtues as he sees them. A woman's propensity to fear was not something to be held against her; it was a fact of *nature*.

It is also worth considering that, although what we read about

Athenian women is viewed through the eyes of men, it is quite probable that in reality women did to some degree conform to masculine expectations. Cultural roles not only disguise themselves as 'nature'; within the history of human society they *become* people's nature. Masculine 'prejudices' about women could appeal to the facts, for the 'facts' had been predetermined by society. In Demosthenes 54, the young speaker, who is bringing a charge of assault, explains to the jurors that:

> when my bearers got to the door, my mother and the women slaves began shrieking and wailing, and it was with difficulty that at length I was carried to a bath.
>
> I was carried home on a litter to the house which I had left strong and well, and my mother rushed out, and the women set up such a shrieking and wailing, as if someone had died, that some of the neighbouring women sent to enquire what had happened.
>
> (Demosthenes 54 [*Konon*]: 9 and 20; Loeb trans., modified)

The young man in question is making no particular point about women, still less complaining about their actions. Their lamentations and panic are adduced merely as part of a credible vignette to illustrate the gravity of the outrage he has suffered. This is the way women behave in crises and can be taken for granted. His only point is that there *was* a crisis which the women's panic thus goes to prove.

When women's lack of fortitude or tendency to indulge in grief, sorrow, or lamentation were criticized it was in those cases where the 'natural' characteristics of women could not be accommodated – or because a *man* was seen to be behaving 'like a woman'. Thus, in the aftermath of Klytaimnestra's murder, Orestes turns on Elektra and rebukes her for her 'womanish lamentations' (Euripides, *Orestes*: 1022ff.) since it is hardly the time for remorse. Similarly, in what does amount to a truly misogynistic outburst, Eteokles attempts to silence the Theban women's lamentations, in Aischylos' *Seven Against Thebes*, but he does so because they are spreading panic throughout the city. In Sophokles' *Trachiniai* the great hero Herakles cries out in his agony:

> Pity me,
> for I seem pitiful to many others, crying

> and sobbing like a girl, and no one could ever say
> that he had seen this man act like that before.
> Always without a groan I followed my painful course.
> Now in my misery I am discovered a woman.
> (Sophokles, *Trachiniai*: 1070–5; M. Jameson trans.)

Herakles certainly invokes the inferior nature of women; but he invokes it because he, a man, has fallen from masculine grace.

In fact women's 'natural' tendency to display grief, sorrow, or fear was not only expected and accepted: it was institutionalized in the ritual mourning which accompanied death. In what might be called an emotional division of labour the order of a male-dominated society virtually demanded that women should exhibit those characteristics thought 'natural' to their sex – the very characteristics which simultaneously proved feminine inferiority. Margaret Alexiou reconstructs the situation in *The Ritual Lament* (1974) from the funeral vases and plaques that illustrate the *prothesis*, the formal lamentation preceding the funeral procession. The father waits at some distance to greet the arriving guests; the kinswomen stand around the bier, with the chief mourner, either mother or wife, at the head, the others behind. Other women, possibly professional mourners, are sometimes grouped on the other side, but it is rare to find men, unless they are close relatives such as a father, brother, or son:

> The ritual formality of the men, who enter in procession usually from the right with the right arm raised in a uniform gesture, contrasts sharply with the wild ecstasy of the women, who stand round the bier in varying attitudes and postures. The chief mourner usually clasps the head of the dead man with both hands, while the others may try to touch his hand, their own right hand stretched over him. Most frequently both hands are raised above the head, sometimes beating the head and visibly pulling at their loosened hair. One painting shows the hair actually coming out
>
> (Alexiou 1974: 6)

There is little doubt that in these cases the women's behaviour was a formalized requirement of the funeral rite (especially if professional mourners were employed). If Athenian society could portray women as incapable of endurance and fortitude in the face of misfortune –

thereby marking them as inferior to men – it could also utilize these failings in order to leave men free to guard their manliness. Women rent their clothes, their hair, and wailed. Men retained a dignified composure, the necessary tears being shed by others.

In this context it should also be remembered that *andreia*, the normal word in Greek for what we would call bravery, courage, or fortitude, was derived from *aner*, the sex-specific term for man, and is more correctly translated as 'manliness'. To credit a woman with the quality of bravery was thus in some ways automatically paradoxical. Not that the paradox remained unexploited. It is used humorously in Aristophanes' *Ekklesiazusai*, when Praxagora compliments her female revolutionaries (who have taken over the citizen assembly) with the words: 'There in the uproar and danger you showed yourself *andreiotatai*,' i.e. 'most manly/brave', the word being complete with its feminine plural termination. Rather more seriously the speaker of the pseudo-Demosthenic *Funeral Oration* proclaims that:

> The Leontidai [the members of one of the ten Athenian tribes] had heard stories about the daughters of Leon and how they had offered themselves to the citizens as a sacrifice on behalf of their country. When these women had such *andreia* [bravery/manliness] it seemed improper to the Leontidai if they, being *andres* [men] should seem inferior to them.
>
> (Demosthenes 60 [*Funeral Oration*]: 29)

But in general (and the above examples do not really constitute exceptions) not only the beliefs of Athenian society, but the very language it used, pre-empted bravery for the male.

III

> Go now. I shall prolong to heaven
> the lamentations and moans and tears
> by which I am overwhelmed forever.
> For it is woman's inborn pleasure always to have
> her present ills upon her lips and tongue.
>
> (Euripides, *Andromache*: 90–5)

Thus speaks (or is made to speak) Euripides' Andromache. But, in the cliché manipulated on the Athenian stage, self-pity, tears, lamentation, were not the only forms of feminine self-indulgence and excess. Two others are notorious, though they are more the stuff of comedy than tragedy: women's fondness for sex, and women's fondness for drink. Both are exploited by Aristophanes with such insistence that he must have been drawing on popular conceptions.

Our own romantic literature tends to subscribe to the view that women are more prone to 'fall in love' than men; or rather, that for women love provides the central focus of their lives while for men it has a more peripheral role. Romance, that is to say, is deemed a feminine preoccupation. To an extent that view is paralleled in Athenian literature. There, too, women are more prone to 'fall in love' than men, and if men are portrayed as subject to the tribulations of Aphrodite, then it is during youth or senility. But our own traditions entail a peculiarity not shared by the Athenians, or not to the same degree: namely, the manner in which our own term 'love', though obviously involving physical attraction, nevertheless seeks to repress, mystify, or even deny its sexual content. Thus while 'romantic' love has been endorsed and encouraged as an honourable female pursuit, the physical aspect of women's 'love' has until recently been heavily veiled.

The Athenian attitude was somewhat different. The sexual referent of *eros* was explicit if not dominant. As Dover (1984: 143) explains, 'affection' as opposed to 'love' was denoted by *philia* (vb *philein*) whose connotations could be non-sexual and which was used of friends (*philoi*) and relatives. True, it can also be translated as 'love'; and as the question 'Do you love me?' whether of a father to a son, or of a youth kissing a girl, employed '*philein*'. But here *philia* is still denoting 'friendship', 'respect', 'kindliness towards'. It is expressive of an attitude rather than an emotion. *Eros* can generate *philia* and vice versa, but it is *eran* which would be most appropriately translated as to 'be in love with', and its cognate *erasthenai* as 'to fall in love with'. Here there is no denial of sexual desire and the portrayal of women as particularly subject to the power of *eros* (or *Eros*, personified and deified) had little to do with a vision of starry-eyed innocence. If women were more prone to 'fall in love' than men, it was because they were particularly subject to their physical desires (cf. Dover 1974: 101).

In Euripides' *Troiades*, Andromache, though by no means endorsing the sentiment, relates that:

> one night, they say, undoes the hatred of a woman to her husband's bed.
>
> (Euripides, *Trojan Women*: 665–6)

And when in Aristophanes' *Clouds*, Right Logic declares that Peleus married Thetis due to his 'chastity' (*dia to sophronein*), Wrong Logic acidly points out that:

> Yes, and she went off and left him too. He wasn't an energetic or sweet fellow to spend the night in bed with. A woman enjoys a rough-and-tumble.
>
> (Aristophanes, *Clouds*: 1068–70)

As Dover (1974: 101) summarizes, 'The Greeks were inclined to think that women desired and enjoyed sexual intercourse more than men.' A famous passage from Hesiod (fr. 275) relates the story in which Zeus and Hera ask Tiresias (who had been both man and woman) which sex got the most enjoyment from sexual intercourse. The answer was that men got one-tenth, and women nine.[3]

It is, however, in the three Aristophanic comedies in which women play the major role – *Lysistrata*, the *Thesmophoriazusai* and the *Ekklesiazusai* – that women's sexual propensities are most prominently portrayed. The plays are rife with allusions to women's adulteries, secreted lovers, and sexual appetites. Sex, of course, has a special prominence in comedy. To see Aristophanic comedy as *directly* documenting the social mores of Athenian society would be absurd. Exaggeration and imagination must play a large part in the humour. But whether or not Athenian women were obsessed with sex, their representations in comedy as being so deserve serious attention. And it should be noted that the humour derived from female sexual exploits in these comedies is based on more than repeated allusions to secreted lovers and betrayed husbands, that is to say, on ridicule aimed at men; it concentrates on the sexual avidity of women themselves.

In the *Lysistrata* the women of Greece declare a sexual strike to force their men into ending the Peloponnesian War. But Aristophanes is at pains to show that the weapon the women employ is double-edged, and the humour of the play derives as much from the discomfiture the strike causes women. In fact one of their initial grievances against the war was that it deprived them of their sexual satisfaction (99–112). Even dildos were in short supply. And when

Lysistrata announced her design to the assembled women, their reaction is scarcely promising:

Lysistrata:	What we must do is to abstain from penises. Why are you turning away from me? Where are you slinking off to? Why are you going pale? What are these tears? Will you do it or not? Tell me.
Myrrine:	I couldn't do it. Let the war go on.
Kalonike:	My god, me neither. Let the war go on.
L:	What about you, little flounder? You said you'd split yourself in two for peace?
K:	Anything, anything you want. I'd walk through fire if I have to. But not penises. There's nothing like them, Lysistrata.
L:	And you?
M:	I'd rather walk through fire.
L:	Oh, what a thoroughly buggered race we are. No wonder they write tragedies about us.

(Aristophanes, *Lysistrata*: 124–37)

Then, with the strike underway, Lysistrata finds difficulty in maintaining solidarity – as she explains in a magnificent parody of tragic diction:

Chorus-Leader:	Queen of our great design and enterprise, tell me, why hast thou come so frowning forth?
L:	The female heart and deeds of evil women cause me to walk despondent up and down.
C-L:	What say'st thou?
L:	Truth! Truth!
C-L:	What is amiss? Tell it to us who love thee.
L:	'Tis shame to speak and grievous to be silent.
C-L:	Do not, I beg, conceal the ill we suffer.
L:	We need a fuck. Such is my tale in brief.
C-L:	Ah! Zeus!
L:	Cry'st thou on Zeus. Anyway, that's how it is. So I can't keep them any longer away from their men. They're slipping off in all directions.

(Aristophanes, *Lysistrata*, 706–19; K. J. Dover trans., 1972: 5)

But an even more startling portrayal of women's sexual avidity is to be found in the *Ekklesiazusai* where the women, disgusted by the mismanagement of the democracy, take over the state. The high point of the play's sexual humour comes when certain consequences of the women's reforms – which entail the abolition of marriage and the institution of a form of economic and sexual communism – are worked out on stage. A young girl and an old hag await their respective lovers. The girl's young man arrives, but before he can enter her door he is seized by the hag who demands him in accordance with the rules of the new constitution. The girl manages to free her lover, but is immediately put to flight by the arrival of a second hag who also lays claim to the youth's sexual services. The second hag is in the process of dragging the youth indoors when a third and yet more hideous crone arrives. The scene ends with the young man being taken screaming inside the house by both old women who seem determined to have their way with him. The abolition of marriage and private property presented in the *Ekklesiazusai* bears a significant resemblance to the proposals put forward by Plato and so strongly criticized by Aristotle.[4] But Aristophanes' comments on the idea are in happier vein: he simply envisages what would happen to men once marriage was abolished and women free – the male would be at the mercy of the predatory female, carried off to satisfy every old woman's lust.

From a strictly modern point of view the Athenians' ample recognition of female libido might appear enlightened. But if women were considered to enjoy sex quite as much as if not more than men, the nature of their sexual appetite was still different, marking them both as inferior, and as inferior in a quite particular way. In Euripides' *Hippolytos*, Theseus angrily abuses Hippolytos for his alleged adultery with Phaidra:

> Are you going to say that 'Foolishness is not to be found in men, but that it is inborn in women'? I know young men who are no better than women in defending themselves against the assaults of Aphrodite on the heart of youth. But they are male, and that helps them.
>
> (Euripides, *Hippolytos*: 966–70; K. J. Dover trans., 1974: 101, extended)

while Sokrates comments that:

> a youth does not share in the pleasures of intercourse with a man as a woman does, but looks on sober as a spectator of

> Aphrodite. Consequently it does not excite any surprise if contempt for the lover is engendered in him.
> (Xenophon, *Symposion*: 8. 21–2; Loeb trans., modified)

The difference between a man's and a woman's sexual desire lay in the fact that a woman was unable to subordinate hers to rational control. Unlike a man, who could choose either to indulge or abstain if needs be, a woman was the slave of her sexual desires.

The Aristophanic portrayal of feminine sexual avidity is closely allied to women's other egregious comic vice: dipsomania. As Lysistrata plaintively remarks in her opening lines:

> Now if they'd been summoned to some shrine of Bacchos, Pan,
> Kolias, Genetyllis, there'd have been no room to move, so
> thick the crowd of timbrels. But now – not a woman to be seen.
> (Aristophanes, *Lysistrata*: 1–4; Loeb trans., modified)

The first lines of the play thus specify wine and sex as women's twin concerns. Aristophanes' portrayal of women as habitual tipplers is constant, but it is in the *Thesmophoriazusai* that this aspect of their character comes to the fore. Disguised as a woman, a certain Mnesilochos has infiltrated the *Thesmophoria*, the women's festival from which men are excluded. The women's suspicions are aroused, and they interrogate the 'matron' to see if she knows the content of their secret rites:

> *Woman*: . . . you, tell me
> what we did first as part of the holy rites?
> *Mnesilochos*: Err, let me see, what was it first? – we drank.
> *W*: Well, what did we do second after that?
> *M*: We drank.
> *W*: Somebody's been letting on to you? What did we do third?
> *M*: Xenylla demanded the drinking-bowl. There wasn't a piss-pot.
> (Aristophanes, *Thesmophoriazusai*: 627–33)

Mnesilochos' (unsuccessful) improvisation no doubt delighted the men of Aristophanes' audience who were probably in ignorance of what precisely did take place during the *Thesmophoria* (as are we). Aristophanes provides a happy answer in accordance with their prejudices – the sacred women's festival is no more than an excuse for Athens' respectable wives to go on a binge. Interestingly, the scene is virtually inverted in the *Ekklesiazusai* where the women, disguised as men, prepare to infiltrate the assembly. Rehearsing the part she must play, and unsure as to correct assembly procedure, the first woman who stands up to speak commences by draining off a bowl of wine. The women's leader, Praxagora, indignantly chides her for thereby giving away the game. But the woman objects:

> What, don't *men* drink in the *ekklesia*. . . . Well, they certainly pour libations, and what's the point of all those tedious prayers unless they get some wine to go with them?
>
> (Aristophanes, *Ekklesiazusai*: 135–42)

Such is women's sad inability to grasp the seriousness of male protocol. To return to the *Thesmophoriazusai*: Mnesilochos, having been discovered, seizes one of the women's babies as hostage in order to defend himself against the outraged matrons – only to find that he is in possession of a wine-skin wrapped in swaddling clothes. His subsequent 'murder' of the wine-skin and the spilling of its 'blood' nevertheless creates quite as much consternation among the women as would the killing of an actual child. In this extended piece of burlesque, the primary female virtue, mother love, is neatly matched with the standard female vice, dipsomania.

Both women's fondness for sex and for wine can, however, be seen as part of a more general portrayal of women as creatures ruled by their physical appetites. Admittedly, the gustatory pleasures are standard motivations (or vices) for most Aristophanic characters, but reference to female indulgence appears as an immediately recognisable source of humour. When Lysistrata muses on the destruction of the Boeotians (along with the other Peloponnesian allies) her companion replies in panic, 'Hey, not *all* the Boeotians; save the eels!' – referring, in a characteristic throw-away joke, to the delicacy of the eels of the lake Kopais. Women's habit of slyly buying themselves dainties 'as of old' is included by Praxagora in the *Ekklesiazusai* as one of women's many and commendable 'conservative' tendencies – a list which in fact runs through the gamut of their stock-in-trade vices. And in the *Thesmophoriazusai*, when one of the women indignantly lists all the little crimes (including adultery and murder) which women can no longer perpetrate since the evil Euripides has exposed them on the tragic stage, she sorrowfully complains:

> We might have put up with that [the exposure of other crimes]
> But, O, my friends, our special little perquisites.
> The corn, the wine, the oil, all gone, gone forever.
>
> (Aristophanes, *Thesmophoriazusai*: 418–20)

Nor is it in comedy alone that gluttony appears as a female failing. In Xenophon's *Oikonomikos* Ischomachos proudly lists the virtues of his new young wife. High among them he places the fact that she has been 'well trained with respect to her stomach'. Admittedly he states that this is a virtue of paramount importance for both men and women, but the implication of his pride is that for women gluttony was an expected vice. Certainly the charge was historically widespread. It appears in Menander's *Dyskolos* and, if the tradition is traced back, then in Hesiod (*Theogony* 599) woman is a voracious 'stomach', a *gaster*, inflicted on man to devour the fruits of his toil; a belly whose voracity, it should be noted, is sexual as well alimentary (Vernant 1981a: 51).

At this stage it should come as little surprise to find that, consonant with their portrayal as subject to all emotions and the desires of the body, women were also considered to lack rationality. The coupling of women's lack of intelligence with their bodily appetites (indeed, with their animality) is nicely illustrated in Aristophanes' *Acharnians*. In an extended series of verbal plays on the word *choiros* meaning both 'pig' and the female sexual organ (729ff.), the Megarian pig-seller tries to pass off his two daughters to Dikaiopolis as a pair of piglets. Explaining the deal in an aside to his daughters, he says:

> Now listen, attend me with your – stomach.
>
> (Aristophanes, *Acharnians*: 733)

The last word is a comic substitution for '*nous*', 'mind', the expected completion of the common phrase. It is to his daughters' stomachs that the Megarian must appeal to show the wisdom of his transaction.[5]

Throughout Athenian literature the assumption that women lacked understanding, foresight, rationality, that they were intellectually inferior to the male, was commonplace. Indeed, the exceptions prove the rule (Dover 1974: 99–100). In Xenophon's *Oikonomikos* (3.11) Ischomachos has doubts as to the responsibility of women as moral agents on the grounds of their innate lack of understanding, which he expresses to Sokrates. Sokrates himself, however, appears to have been something of a free thinker on the subject, as his views in Xenophon's *Symposion* (2.8–12) bear witness – but in a most revealing manner:

> Gentlemen, in many other things it is clear, and in those things which this slave is doing [juggling and acrobatics], that woman's nature is not at all inferior to man's – except in that it lacks understanding and strength. So if any of you has a wife, let him confidently set about teaching her whatever he would like to have her know.
>
> Witnesses of this feat [the slave-girl diving through hoops of swords] will never again deny, I feel sure, that courage (*andreia*) admits of being taught, when this girl in spite of being a female leaps so boldly in among the swords.
>
> (Xenophon, *Symposion*: 2.9–12; Loeb trans., modified)

In fact the only point that Sokrates seems really to make is that even those who innately lack understanding (*gnome*) can still be trained to virtue, and where, in tragedy, certain women are manifestly equipped with intelligence and foresight, then this becomes an idiosyncratic abnormality. Klytaimnestra is referred to as 'man-minded' in Aischylos' *Agamemnon* because she demonstrates a superior intellect – an attribute which, under the circumstances, was not entirely complimentary (Pomeroy 1975: 98).[6] And to the chorus of Argive elders, slow to digest the fact that she has murdered Agememnon, she impatiently retorts:

> You are examining me as if I was a senseless woman.
>
> (Aischylos, *Agamemnon*: 1401)

It is the tenacity of the commonplace conception of women as lacking intelligence that elevates Klytaimnestra into such an extraordinary figure, for as Dover summarizes, 'A woman, in fact, was thought to have a "butterfly" mind, equally incapable of intelligent, farsighted deliberation and of foregoing the emotional reaction of the moment in pursuit of distant and impersonal aims' (1974: 100).

IV

Emotional, self-indulgent, inebriate, gluttonous, irrational, weak-willed: these form part of the commonplace description of women in the writings of fifth- and fourth-century Athens. They are not perhaps the accusations which every individual man levelled against his own wife, mother, daughter, or sister; but they are the substance of a stereotypical image which surfaces across the range of Athenian texts, tragic, comic and, as will be seen, forensic and even philosophical. Admittedly their echoes are discernible within our own traditions. The differential attributions of 'bravery' and 'fear' to men and to women, or of endurance and emotional self-indulgence, or even of reserve and loquacity, are a familiar feature of male discourse to this day. Nevertheless, as I hope to show, such characterizations and such contrasts did form part of a culturally specific ideological system, for within Athenian thought the commonplace characterizations of women related to a more general opposition between those who possessed the primary virtue of self-control and those who did not. It is this opposition which is crucial, for on it turn the Athenian notions of freedom and subordination, notions themselves grounded in Athens' economic structure, in the fact that it was a slave-owning society. And here, of course, is the nexus between politics and the attributes of gender; it is the opposition between those innately possessed of self-control, and those who lack it, that ideologically renders women's subordinated place within the social structure of the *polis* a 'natural' one.

The degree to which Athens' economy was dependent on slave-labour remains a vexed issue. This does not alter the fact that Athens *was* a slave-owing society; that its whole character was permeated by slavery. Indeed, what was distinctive about Athenian slavery was not the existence of a body of people constrained to labour for others, nor the size of that body of labourers, nor even the conditions under which they laboured, but the radical opposition created between two clearly defined groups: those who were wholly free, autonomous, and (within the limits of a law in any case of their own devising) answerable only unto themselves, and those who, whatever the practical conditions of their treatment, were wholly slave, without any rights of self-determination and subject to the will of those whose chattel-property they were.[7] Such an opposition owes as much to the development of democracy in Athens as it does to any deterioration in the conditions of servitude.[8]

But if 'free' (*eleutheroi*) and 'slave' (*douloi*) referred to two clearly defined categories of person within the Athenian social structure, the ideological concomitants of freedom and slavery were far more extensive and far less well-defined. Plato's *Republic* was a theoretical work, but it exemplifies a mentality in which the opposition between free and slave has become a basic mode of classification. As Joseph Vogt summarizes in his *Ancient Slavery and the Ideal of Man*:

> The contrast between free man and slave, like the tension between father and son, magistrate and citizen, corresponds to differences in the essential value of particular human beings and in the functions of various social groups. All who lack *logos* (reason) are *douloi* (slaves); yet the notion of *douleia* (slavery) is also used to describe any kind of subjection, whether to the government, to parents, or to the rule of law, and this willingness to obey is considered beneficial to the subject. The body is the slave of the soul, and this relationship between ruling and serving applied beyond the limits of the human world to the cosmic world. A natural law of hierarchy applies to the structure of the soul, to relations between status groups and to the qualities of the cosmic forces behind them.
>
> (1974: 33)

The use of 'slave' and 'free' to describe a variety of social relationships, individual conditions, or even personal characteristics beyond the actual social institution of slavery was not, however, confined to Plato or to philosophical writings. The opposition between slave and free informed Athenian thought as a whole, and though obviously born of the existence of institutionalized slavery it did not necessarily have direct reference to the legally defined categories of slave and free.

Here it is important to note that however large a part slave-labour may have played in the economy of Athens, slavery did not replace free labour but coexisted with it. What 'slavery' frequently referred to in its general or extended sense was not therefore the necessity to undertake labour itself, but *any external compulsion* which deprived a man of his self-determination, autonomy, and independence. Any man who, willingly or unwillingly, was placed in a position where he was subject to somebody else could be said to be acting 'slavishly'. The slavishness resided not in what he was required to do, but in the fact that he was required to do it for another (cf. Dover 1974: 40).

The situation can be illustrated by the following contrast. The life of the Attic peasant farmer was doubtless a hard one yet, in the literature of Athens, the peasant farmer enjoys considerable respect. In Aristophanes he is the hero of his time, and it was in hard work that his virtues were seen to lie (Ehrenberg 1974a:73; Dover 1974: 113). Even Xenophon's gentleman farmer, Ischomachos, can praise industry, and in himself as well as for others. On the other hand, the orator Demosthenes could make considerable capital from the fact that his rival, Aischines, had received *wages* in a number of capacities; from the fact that he had been *employed* by others:

> You taught reading and writing; but I went to school. You initiated; but I was initiated. You were secretary; but I was a voting member. You played third parts; but I was in the audience.
>
> (Demosthenes 18 [*On the Crown*]: 265)

Demosthenes could not have been arguing that Aischines' employments were in themselves arduous or unpleasant. Nor could he have been trying to suggest (before a court drawn from the common people) that any Athenian citizen ought to live a life of constant leisure. What he does suggest is that Aischines has lost his moral integrity by having placed himself at the hired disposal of others; that he has become no more than a bought man (cf. Dover 1974: 32–5).

Similarly, in Xenophon's *Memorabilia*, Eutheros, having fallen on hard times, is already 'working with his body'. But his response to Sokrates' suggestion that he should seek an easier occupation by becoming some land-owner's bailiff is, 'I simply couldn't stand being a slave' (2.8.4). As bailiff, Eutheros would have been managing slaves, not legally selling himself into slavery. But this is not the point. The slavery Eutheros fears is a loss of personal autonomy.[9] As Aristotle expresses the matter in the *Rhetoric* (1367a33), 'The condition of a free man is that he does not live for the benefit of another.' And just how far this attitude could be carried (admittedly by a somewhat exceptional character) is illustrated in Xenophon's description of Sokrates himself:

> For while he (Sokrates) checked their other desires he would not make money out of their desire for his companionship. He thought that by holding himself back from this he was safeguarding his freedom (*eleutheria*). Those who charged a fee for their company he denounced as selling themselves into slavery, since they were compelled to converse with whoever paid the fee.
>
> (Xenophon, *Memorabilia*: 1.2. 5–6; Loeb trans., modified)

Here the contrast between freedom and slavery has clearly been transposed from both a question of labour and a question of defined legal status to a matter of self-estimation and personal moral integrity (cf. Dover 1974: 114–16, 171–2).

In this context it actually becomes questionable as to which was conceived to be prior: the socially and legally defined categories of slave and free, or the type of behaviour or even attributes of 'character' which were thought befitting a citizen and a free man and conversely a slave.[10] Aristotle, who so often expresses as an explicit generalization the commonplace assumptions of his time, is willing to argue that there are in fact two types of slavery and of slave: those who are slaves by nature, and those who are legally enslaved (*Politics*: 1254a13–1255b15). Ideally the two types ought to coincide and Aristotle, with some qualification, thinks that for the most part they do. It is on his notion of 'natural' slavery that his notorious justification of institutionalized slavery rests (ibid., 1253b14ff.). Slaves, although human beings, are nevertheless naturally slavish because for the most part the deliberative faculty of their psyche is lacking. The institution thus accommodated those who were in any case 'psychologically' incapable of self-determination.[11] They, *par excellence*, were barbarian non-Greeks.[12] But the attributes of character, the manifestations of behaviour, which marked a man as 'unfree', as *aneleutheros*, could still be found among the citizen body and among those who were neither slaves nor barbarians. 'Freedom' as an attribute of character, as indeed a moral virtue, was an ideal to be maintained by the citizenry – and like all forms of virtue it was under the continual threat of vice.

V

Let me now pursue a slightly different path in tracing the ideological concomitants of freedom and slavery.

As Dover remarks, there has been a widespread belief that 'the Greeks lived in a rosy haze of uninhibited sexuality, untroubled by the fear, guilt, and shame which later cultures were to invent' (1974: 205). It should already be apparent that such a belief is largely romantic nonsense. Nevertheless it is true that Athenian attitudes towards the bodily enjoyments and the integration of the erotic into public life are in contrast with the mores of later (Christianized) western cultures. Old Comedy, performed publicly and as part of the official religious life of the *polis*, was bawdy in the extreme. Aristophanes' linguistic inventiveness raised obscenity to a fine art while the extended phalli worn by comic actors could be turned to provide any number of visual jests. Athenian decorated pottery depicted almost every variety of human sexual exploit. As Dover points out, the Greeks did, after all, treat sexual enjoyment as the province of a goddess, Aphrodite, calling it 'what belongs to Aphrodite', *ta aphrodisia* (1974: 205). And though, as Dover argues, there was a growing prudery in the Athens of the fourth century and a tendency to confine the sexual to privileged areas of artistic expression, giant models of the erect penis were still carried in procession during the Dionysian festival and phalli adorned the *herms* positioned on street corners and in front of private homes. Even in law-court oratory there was a considerable frankness about enjoyment of the sensual life. If chastity was a virtue, it was a virtue reserved for the wives and daughters of Athenian citizens. As already seen, the male Athenian could satisfy his sexual desires quite freely whether in the brothels of Athens, more elegantly in the company of kept *hetairai*, with his own female domestic slaves, at male drinking parties (*symposia*) with hired sexual entertainers, and finally (though in accordance with more complex conventions) in homosexual pursuits. Some elementary rules of social propriety might be observed with respect to the feelings of a legitimate wife, but the freedom of the Athenian citizen to pursue a wide variety of sexual interests seems to have been an accepted right.

For all that, the writings of philosophers and moralists show an extreme wariness, if not open hostility, towards sexual pleasure and towards the emotions and desires it aroused. Xenophon's Antisthenes, Sokrates and Prodikos (*Symposion*: 4.38 and *Memorabilia*: 1.3.14 and 2.1.30) may have been taking an extreme point of view when they suggested that the good man ought either to forego sexual intercourse altogether or else satisfy himself as quickly and cheaply as possible by masturbation, but a negative attitude towards sexuality is characteristic of philosophical writings (Dover 1974: 211, 213). Nor is there much doubt that popular sentiment was also hostile towards excessive sexual indulgence – an indulgence which, it should be noted, was classed along with excessive expenditure on food, drink, clothing, or on any other bodily comfort or pleasure (Dover 1974: 179–80, 208–9). Demosthenes, for example, could abuse an opponent with the words:

> He went around purchasing prostitutes and fish.
>
> (Demosthenes, 19 [*Embassy*]: 229)

As for the orators, their speeches are full of attacks on the incontinence of their opponents. To take but one example from Aischines' massive diatribe against Timarchos:

> For his father had left him a very large estate which he squandered, as I shall show in my speech. But he behaved as he did because he was enslaved [*douleuon*] by the most shameful pleasures: by gluttony and extravagant dinner-parties and flute-girls and *hetairai* and dice and all the other things by which no properly born and free man should be mastered.
>
> (Aischines 1 [*Timarchos*]: 42; Loeb trans., modified)

Athenian attitudes towards the physical pleasures were, then, ambivalent; but they were not contradictory. There is a consistent basis to the moral evaluations which becomes clearer if the language is examined. Aischines claimed that Timarchos' behaviour resulted from his 'enslavement' to pleasures which mastered him in a way unbefitting a man who was 'properly born' and 'free'. In Lysias 21 [*Defence against a charge of taking bribes*] the speaker appeals to the court to:

> think also of my private conduct, reflecting that this is the most laborious of services, to be *ordered and self-controlled* and neither *to be worsted by pleasure* nor incited by prospect of gain.
>
> (Lysias: 21.19)

In Demosthenes 40 [*Boiotos II*] the speaker explains:

> However, he was not so wholly *conquered by his desire* as to deem it right even after my mother's death to receive this woman into the house.
>
> (Demosthenes 40 [*Boiotos II*]: 9; Loeb trans., modified)

And to take one of the many examples from Xenophon's writings, Sokrates explains:

> Yes, these too are slaves (*douloi*), and hard indeed are their masters. Some are in bondage to gluttony, some to lechery, some to drink, and some to foolish and costly ambitions. And so hard is the rule of these, that so long as they see that he is strong and capable of work, they force him to pay over all the profits of his toil, and to spend it on their own desires; but no sooner do they find that he is too old to work, than they leave him to an old age of misery and try to fasten their yoke on the shoulders of others. Ah, Kritoboulos, we must fight for our *freedom* against them as persistently as if they were armed men trying to *enslave* us. Indeed, open enemies may be gentlemen, and when they enslave us, may, by chastening, purge us of our faults and cause us to live better lives in future. But such mistresses as these never cease to plague men in body and soul and property all the time that they have power over them.
>
> (Xenophon, *Oikonomikos*: 1.22–3; Loeb trans., modified)

The recurrent metaphor (if metaphor is not too weak a word) is of *slavery and subordination* to the desires which are aroused by physical pleasure;[13] of a loss of freedom attendant on physical indulgence. As Xenophon's Sokrates elsewhere explains, by means of one of his characteristic interrogations:

> Do you think that freedom is a noble and splendid possession both for individual and for communities?
> (Yes)
> Then do you think the man is free who is ruled by bodily pleasures and is unable to do what is best because of them?
> (No) . . .
> You feel sure, then, that those who are without self-control are unfree.
>
> (Xenophon, *Memorabilia*: 4.5.2–4; Loeb trans., modified)

But it should be noted that their physical pleasures are still not denied as being in themselves legitimate enjoyments – and even by someone so extreme as Sokrates. For he goes on to say:

> Incontinence will not let them endure hunger or thirst or sexual desire or lack of sleep, which are the sole causes of pleasure in eating and drinking and sexual indulgence, and in resting and sleeping after a time of waiting and resistance until the moment comes *when these will give you the greatest possible satisfaction.*
>
> (ibid., 4.5.9)

In short, what appears to have been considered reprehensible was not so much pleasure itself – whether of food, wine, or sex – but any excessive indulgence in pleasure which might result in 'enslavement' to it. The idea, for example, that the good man should forego sexual intercourse or else satisfy it as quickly and cheaply as possible cannot be taken as part of a moral system within which the bodily desires were seen as intrinsically 'evil'. And, as Dover notes, 'Xenophon's Sokrates, although hostile to the body on such issues as appeared to involve a clear mind/body antithesis, lists among the blessings conferred on mankind by beneficient providence (Xen., *Mem.* I iv 12) the fact that we, unlike so many animals, can be sexually active in all seasons' (1974: 205).

For the moralists, the bodily desires were *potential* evils, reprehensible not in themselves, but in what they might lead to – a loss of freedom, of 'self-control', of 'autonomy'; in short, the enslavement of the mind by the body. And though Aischines in the above quoted passage does talk of 'shameful pleasures', what he is primarily concerned to prove is not that his rival, Timarchos, had a taste for food and sex, but that Timarchos' indulgence in these pleasures was so great that he was incapable of directing his actions towards any other goal. They had become his masters; he their slave. Who could then trust such a man? He had forfeited his independence, his ability to determine what was right and good, and his ability to pursue what a free man should.

In this context it is worth examining Aischines' politically motivated prosecution of Timarchos in a little more detail.[14] A lurid description of Timarchos' libertine career forms a large part of Aischines' speech:

> But when these resources had been wasted and gambled away and eaten up, and the defendant had lost his youthful charm,

> and, as you would expect, no one would any longer give him anything, while his disgusting and wicked nature (*physis*) constantly craved the same indulgences, and with excessive incontinence kept making demand after demand on him, then, at last, incessantly drawn back to his old habits, he resorted to devouring his patrimony. And not only did he eat it up, but, as it were, he drank it up.
>
> (Aischines 1 [*Timarchos*]: 95; Lob trans., modified)

> In place of his patrimony, the resources he left were lewdness (*bdeluria*), calumny (*sykophantia*), impudence (*thrasos*), cowardice (*deilia*), unmanliness (*anandria*), a face that knows not the blush of shame – all that would produce the lowest and most unprofitable citizen.
>
> (ibid., 105)

The passages could well have been pronounced by Sokrates. Incontinence leads straight to moral degeneration. Indeed, as Xenophon's Sokrates does put it:

> men who take to drink or get involved in love intrigues lose the power of caring what is right or wrong, and of avoiding evil. For many who are careful with their money no sooner fall in love than they waste it; and when they have spent it all, they no longer shrink from making more by methods which they formerly avoided because they thought them disgraceful.
>
> (Xenophon, *Memorabilia*: 1.2.22; Loeb trans., modified)

> Of sensual passion Sokrates would say: 'Avoid it resolutely: it is not easy to control yourself once you meddle with that sort of thing. . . . Poor fellow! What do you think will happen to you through kissing a pretty face? Won't you lose your liberty in a trice and become a slave (*doulos*), begin spending large sums on harmful pleasures, have not time to give to anything fit for a gentleman, be forced to concern yourself with things that no madman even would care about?'
>
> (ibid., 1.3.8)

It should nevertheless be noted that the actual charge on which Aischines prosecutes Timarchos was more specific: a charge based on two laws attributed to Solon and paraphrased in Aischines' speech. In essence they ordained that any citizen who prostituted himself was to suffer total *atimia*, that is to say, he was to be deprived of all civic rights. These laws (and related ones concerning the prostitution of an Athenian youth by his parent or guardian) are of particular interest.

As we have seen, homosexual relationships between male Athenians were common, perhaps even normal, and as Dover has stressed, it would be foolish to think that such relationships were always 'Platonic'. Indeed, what makes the treatment of homosexuality by Plato and other moralists so fascinating is precisely their assumption that it was in a homosexual (rather than a heterosexual) context that desire (*eros*) would be aroused, and then their consequent attempts to moralize that desire and to accommodate it to the requirements of an ethical system. The desires of the body and the emotions they aroused were dangerous in that they threatened enslavement of the *psyche*; but homosexual *eros* could be rendered virtuous by the spiritual relationship it engendered between two men constantly vying to gain each other's esteem – and thus by the *resistance* they were required to cultivate against physical desire. Homosexual *eros* becomes actually conducive of virtue by reinforcing self-control. One might be tempted to think this argument a little specious, but it was not the philosophers alone who resorted to it. In his prosecution of Timarchos it was vital for Aischines to prove that Timarchos was not simply a 'lover', but a prostitute. Aischines knew only too well that he was in no position to condemn homosexual relationships *tout court*:

> I understand he is going to carry the war into my own territory and ask me if I am not ashamed myself, after having made a nuisance of myself in the gymnasia and having been many times a lover (*erastes*), now to be bringing the practice into reproach and danger. And finally – so I am told – in an attempt to raise a laugh and start silly talk amongst you, he says he is going to exhibit all the erotic poems I have ever addressed to one person or another, and he promises to call witnesses to certain quarrels and fights in which I have been involved in consequence of this habit.
>
> (Aischines 1 [*Timarchos*]: 135; Loeb trans., modified)

Instead, Aischines adopts a stance remarkably similar to the apologies for homosexual eroticism familiar from the writings of Plato,

Xenophon, or the author of the *Erotic Essay* in the Demosthenic corpus:

> The distinction which I draw is this: to be in love with those who are beautiful and prudent (*sophon*) is the experience of a kind persuasion and a righteous mind; but to hire for money and to indulge in licentiousness is the act of a wanton and ill-bred man. And whereas it is an honour to be the object of a pure love, I declare that he who has played the prostitute by inducement of wage has disgraced himself.
>
> (ibid., 137)

And Aischines concludes this passage with an appeal to the court to endorse the distinction he has made, thereby condemning Timarchos:

> To which class do you assign Timarchos – to those who are loved or to those who are prostituted? You see, Timarchos, you are not to be permitted to desert the company which you have chosen and go over to the ways of free men (*eleutheroi*).
>
> (ibid., 159)

Whether Aischines' own version of commendable *eros* was supposed always to be unconsummated is difficult to tell from the deliberately evasive language (cf. Dover 1978: 48); but what is clear is that the only *crime* with which Aischines could charge Timarchos was that he had prostituted himself – that he had accepted money and become a sexual *misthotos* (cf. Dover 1978: 19–34). But what made the Athenian citizen's prostitution of himself so heinous that the law took special note of it and punished it with the severity of *atimia*? It could not have been moral outrage at the idea of homosexuality itself. Rather, it was the concordance of two closely related aspects of the same fundamental notion: that in any erotic relationship one's 'freedom' (autonomy, self-control, moral responsibility) was put at risk; that in any paid employment or hire one's freedom was similarly forfeit. If physical desire was dangerous since it threatened to enslave the individual, sexual desire was particularly so since it could involve not only subordination to the desire itself, but also to another individual.

Athenian morality therefore required that a youth should resist the sexual advances made towards him; even if he yielded, the relationship was still between two free men; and in this complex of emotional and physical demands and attractions exactly who was enslaving whom remained unclear. In the case of the prostitute no such ambiguity existed. He had sold his person, submitting himself entirely to another. He was the slave both of sexual desire and of the man who had bought him. He had demonstrated to the full that he was not a free man; and the state relieved him of the rights accruing to one (cf. Dover 1978: 106–9). As Aischines glosses the law of Solon:

> the man who has violently made trade with his body, he [Solon] thought would also be ready to sell the common interests of the *polis*.
>
> (ibid., 30)

In short, the ideological concomitants of freedom and of slavery, though keyed to the social and economic structure of Athens, were extended to apply to all forms of independence and subordination. Prostitution was perhaps the limiting case of a form of 'slavery' that resulted from both a sale of labour and a submission to sexual desire – and it shows clearly enough that such slavery, though descriptive of moral and behavioural qualities, was not, *even in the case of men*, disconnected with the legal ability to exercise the rights of citizenship. Freedom referred not only to the conditions of one's existence but to the nature of one's existence. Freedom was autonomy; it was personal integrity; it was self-determination; it was *self*-control. And ideally, only those possessed of self-control were worthy of the rights of citizenship – of self-determination in its legal form.

VI

In tracing the Athenian notion of freedom and slavery, one last matter deserves attention. If one could be 'mastered', 'overcome', 'enslaved', 'conquered', not only by other people but also by physical appetites and by the emotions and desires they aroused, it would seem that such appetites, emotions, and desires were thought of as in some way separate from the self. This, however, is dangerous ground, for it is not easy to know to what extent a *façon de parler* can be equated with a *façon de penser* (Dover 1974: 124–6; cf. Dodds 1951: 4–5).

It is clear that concomitant with a wariness of the bodily appetites, desires, and passions was a heavy moral emphasis on the role of reason. 'Nothing in human nature is more important than *logismos*', say Menander (fr. 213), and this was not a profundity the playwright had stumbled on, but a commonplace of his time. The orator Lykourgos, quoting one of the 'old poets', assures his audience that, 'When the anger of the daimons is injuring a man, the first thing is that it takes the good understanding out of his mind' (Dodds 1951:39). And the author of the Demosthenic *Erotic Essay* can moralize by saying:

> Now, of the powers residing in human beings we shall find that intelligence leads all the rest.

Further:

> of all things the most irrational is to be ambitious for wealth, bodily strength . . . all of which are perishable and usually slaves to intelligence.
>
> (Demosthenes 61 [*Erotic Essay*]: 37; Loeb trans., modified)

Similarly progress from infancy to middle-age was seen as a continuous development of the rational faculties (Dover 1974: 102). The young were unfitted to rule and their advice taken lightly since their rationality was in question. So was that of the old. Athenian respect for age did not alter the fact that it was considered to entail a diminution of the rational faculties. Xenophon consoled himself on Sokrates' death with the thought that:

> First he [Sokrates] had already reached such an age that had he not died then, death must have come to him soon after. Secondly, he escaped the most irksome stage of his life and the inevitable diminution of mental powers.
>
> (Xenophon, *Memorabilia*: 4.8.1; Loeb trans., modified)

At the other end of the scale Aristotle (*Nikomachean Ethics*: 1095a 2–9) would not allow the young to study political theory on the grounds of their inexperience and susceptibility to powerful emotions (Dover 1974: 102 n.1). Such sentiments were also embodied in the laws which, on the one hand, reserved the holding of most public offices to men over thirty and, on the other hand, allowed senility as an argument for the invalidation of a will. Indeed, the handing over of an estate by a father to his son on the grounds that he was no longer competent to manage it in his old age was a common practice.

What is not so clear is the extent to which the opposition between the rational as identified with the 'self' and the emotions and passions as identified with something other than and exterior to the self was anything more than linguistic idiom. The Greeks' conception(s) of the self and the meaning of the term '*psyche*' are notoriously problematic (indeed, to arrive at any clear idea of what the 'self' now means in contemporary society would be no mean achievement). On the one hand there are the non-systematic representations of ordinary language and usage from which some psychological model of man can be inferred – but it is open to arguments about the autonomy of linguistic expression, the role of metaphor, and the endurance of literary convention, etc. On the other hand there are the works of professional philosophers, who offer a reasonably lucid account of their conception of the *psyche* – but philosophical opinion does not always reflect the views of society as a whole.

Nevertheless, the tendency in both natural language and in philosophical writing towards the reification and separation of psychological entities, and particularly towards the externalization of the emotions, is remarkably consistent throughout Greek history. The self in Homer appears to lack any unity, and as Dodds writes: 'Homer appears to credit man with a *psyche* only after death, or when he is in the act of fainting or dying or is threatened with death: the only recorded function of the *psyche* in relation to the living man is to leave him. Nor has Homer any other word for the living personality' (1951: 15–16). Homeric man is, however, in possession of other entities, *noos* and *thymos*, which, along with such bodily organs as the *phrenes* (lungs) and *kardia* (heart), seem to determine his actions. The *thymos*, something like an 'organ of feeling', was perhaps the seat of the emotions or perhaps their cause, but in either case Homeric man could converse with his *thymos* (as he could with his heart or his belly) as if it were endowed with some independence, and Dodds feels confident that: 'for Homeric man the *thymos* tends not to be felt as part of the self: it commonly appears as an independent inner voice.' It is to 'this habit of (as we should say) "objectifying emotional drives", treating them as not-self' that Dodds relates his idea of 'psychic interventions' (1951: 16).

Such 'psychic interventions' are *ate* – 'folly', 'madness', 'infatuation', which robs men of their wits; *menos* – a heightened state of daring, courage, strength or exhilaration; or indeed almost any sort of sudden mental occurrence. These 'psychic interventions' are generally attributed to some nameless *daimon*, 'god' or 'gods'. *Noos*, on the other hand, would appear to be the seat of, or to constitute, a man's intellectual or rational faculties. And as Dodds observes, the frequency of references to the *noos* in the Homeric epics is part of a characteristic habit, 'of explaining character or behaviour in terms of knowledge' (1951: 16). As Dodds proceeds to argue:

> This intellectualist approach to the explanation of behaviour set a lasting stamp on the Greek mind: the so-called Socratic paradoxes that 'virtue is knowledge' and that 'no one does wrong on purpose', were no novelties, but an explicit generalized formulation of what had long been an ingrained habit of thought. . . . If character is knowledge, what is not knowledge is no part of the character, but comes to a man from outside. When he acts in a manner contrary to the system of conscious dispositions which he is said to 'know', his action is not properly his own, but has been dictated to him. In other words, unsystematized, non-rational impulses, and the acts resulting from them, tend to be excluded from the self and ascribed to an alien origin.
>
> (1951: 17)

Homer should not be pressed to yield a coherent account of the early Greeks' concept of the 'personality' or 'self' (nor is there any reason to suppose that an explicit concept existed). Moreover, what is found in Homer cannot be equated with ideas of the classical period. Nevertheless, as Dodds observes:

> When Theognis calls hope and fear 'dangerous daemons' or when Sophocles speaks of Eros as a power that 'warps to wrong the righteous mind for its destruction' we should not dismiss this as 'personification': behind it lies the old Homeric feeling that these are not truly part of the self, since they are not within man's conscious control; they are endowed with a life and energy of their own, and so can force a man, as it were from the outside, into conduct foreign to him.
>
> (1951: 41)

Indeed, the representation of the emotions or desires as external agencies affecting man continued to appear throughout classical literature, especially tragedy. Here the representations are of course more elaborate and, it might be argued, more consciously 'literary' or 'allegorical'; but at the same time they are evidence that the externalization of emotional forces remained a mode of representation at least acceptable to commonsense (Padel: forthcoming). The continued reification and personification of emotional forces both in language and, physically, on the stage, are thus of perhaps more than 'poetic' significance. Aischines, in the fourth century, could still address a jury with these words:

> You must not imagine, citizens, that the impulse to wrongdoing comes from the gods. No, rather it is from the wickedness of men. Nor must you think that ungodly men are, as in tragedy, driven and chastised by the Furies with blazing torches in their hands. No, the impetuous desires of the body and their insatiability – these are what fill the robber's band and send men aboard the pirates' boat.
>
> (Aischines 1 [*Timarchos*]: 191; Loeb trans., modified)

Moral responsibility is asserted and the images of the poets denied, but criminality continues to be ascribed to something other than conscious deliberation. It is not so much man's will as his lack of will which is to blame.

In this context, the Athenians' views on insanity are interesting. First, madness (*paranoia, mania*) was not generally considered as a chronic state caused by some inherent or internal malfunctioning of the *psyche*; rather it was a temporary affliction, a sickness, a *nosos* which acted on or deprived a man of his wits. Second, 'the distinction between emotion, thoughtlessness and shamelessness . . . and insanity . . . was generally treated by the Greeks as quantitative' (Dover 1974: 127). Thus acts of criminality or folly could be termed as acts of 'madness' and – although it should be noted this was no excuse before the law – attributed to something external to the reasoning self, for 'any state of mind which is unwelcome or may have bad consequences could be called *nosos*, "sickness", "illness" ' (Dover 1974: 125).

In the more formalized expressions of the philosophers, however, feelings, emotions, passions, desires are not external forces; but whether in Plato's tripartite *psyche* or Aristotle's bipartite *psyche* a separation is still carefully maintained between man's logical and non-logical faculties: between his powers of reasoning and deliberation on the one hand, and his emotions, desires, and bodily appetites on the other. The precise nature of the interrelationship(s) between these faculties in the psychological theories of both Plato and Aristotle is complex and beyond the scope of this work. It is enough to note, however, that a separation was maintained and that what was required for good conduct was the subordination of the emotions and passions to reason.[15] Indeed, Plato's commitment to rationality and his hostility towards the bodily pleasures and the desires which they aroused are constant. The emotions and desires are seen as antagonistic to reason, *logos*, which must prevail if virtue is to be achieved. In the *Phaido* the conflict is between the desires of the body on the one hand and the *psyche* as a whole on the other. But in the *Republic* and the *Sophist* the conflict is internalized to the *psyche* itself, to its antagonistic elements. When, in the *Timaios*, Plato returns to the idea of the pure, transcendent, immortal *psyche*, he does so by the expedient of virtually sundering man in two, and by endowing him with two different *psychai* (Dodds 1951: 212–14):

> And they [the off-spring of the creator], imitating him, received from him the immortal principle of the *psyche*; and around this they proceeded to construct a mortal body, and made it to be the vehicle of the soul, and constructed within the body a *psyche* of another nature which was mortal, subject to terrible and irresistible affections – first of all pleasure, the greatest incitement of evil; then pain, which deters from good; also rashness and fear, two foolish counsellors; anger, hard to be appeased, and hope easily led astray – these they mingled with irrational sense and with all-daring *eros* according to necessary laws, and so framed man. Wherefore, fearing to pollute the divine any more than was necessary, they gave the mortal nature a separate habitation in another part of the body, placing the neck between them. . . .
>
> That part of the inferior *psyche* which is endowed with courage and passion and loves contention they settled nearer the head, midway between the midriff and neck, in order that being obedient to the rule of reason (*logos*) it might join with it in controlling and restraining the desires when they are no longer willing of their own accord to obey the command of reason issuing from the citadel.
>
> (Plato, *Timaios*: 69c–70a; B. Jowett trans., modified)

Aristotle's hostility towards the non-logical side of man's nature is by no means so pronounced as Plato's. For him virtue is acquired not by knowledge or reflection, but through habituation. Virtue, in his famous definition, is,

> a disposition of the *psyche* in which, when it has to choose among actions and feelings, it observes the mean relative to us, this being determined by such a rule or principle as would take shape in the mind of a man of sense or practical wisdom.
> (Aristotle, *Nikomachean Ethics*: 1107a; J. A. K. Thomson trans., modified)

Such a disposition requires the habituation of the passions, and the passions thereby participate directly in the formation of a virtuous man. In fact man is 'rational' in two ways: he is actively and passively rational, and there are accordingly two sides to human virtue. The former is the possession of a *logos* (rationality); the latter is a discipline of passion which is in essential conformity to *logos*. But the priority is with the actively rational principle in the sense that it is for this to determine the end to which the passively rational conforms. Aristotle does not want to negate the non-rational aspects of man, but rather to mould them. Far from being irredeemably dangerous, it is through habituation of the bodily desires and of the passions that a man is first led towards virtuous conduct.

Nevertheless, the distinction between the logical and the non-logical (the emotions, the passions, the bodily desires) is maintained and the priority of *logos* as the determinant of virtue reasserted. A child, a woman, even a slave can, through habituation, acquire the virtues 'relative to them' without the exercise of rational deliberation; but it is still practical wisdom which involves *logos*, reason (whether possessed by the person himself or, as in these cases, by someone in control of that person), which determines the 'mean relative to them', and which thus defines for them the nature of virtuous conduct.[16] And if Aristotle allowed the emotions and passions a prominent place in his ethical theory, according to tradition he was still in no doubt that, untutored, the passions and emotions led straight to chaos and to vice:

> It is said in a certain book of the ancients that the pupils of Aristotle assembled before him one day. And Aristotle said to them: 'While I was standing on a hill I saw a youth, who stood on a terrace roof and recited a poem, the meaning of which

> was: Whoever dies of passionate love, let him die in this manner; there is no good in love without death.' Then said his pupil Issos: 'O philosopher, inform us concerning the essence of love.' And Aristotle replied: 'Love is an impulse which is generated in the heart; when once it is generated, it moves and grows; afterwards it becomes mature. When it has become mature it is joined by affections of appetite whenever the lover in the depth of his heart increases in his excitement, his perseverance, his desire, his concentrations, and his wishes. And this brings him to cupidity and urges him to demands, until it brings him to disquieting grief, continuous sleeplessness, and hopeless passion and sadness and destruction of mind.'
>
> (Al-Dailami, cod. Tubingen Weisweiler 81; Ross trans., 1952: 26)

VII

Let me now try to draw together the threads of this chapter's argument. In the preceding sections I have tried to outline three pairs of oppositions which I consider to be closely related in Athenian thought: freedom/slavery, self-control/incontinence, rationality/the bodily appetites and emotions. The oppositions are not synonymous, but their connotations overlap. Freedom referred not only to a legally defined status, but to a general condition of autonomy; autonomy meant freedom not only from the command of others, but also from the bodily desires and emotions; freedom from the bodily desires entailed 'self-control', the mastery by the rational self of those bodily desires and emotions which threatened to enslave it. As part of a system of values, a mistrust of the emotions, a suppression of the bodily desires, a lauding of the rational, and an emphasis on self-control are not uniquely Greek; nor, within the Athenian context, would it be possible to explain their occurrence simply by reference to the existence of institutionalized slavery. Nevertheless it seems to me that the fact that Athens was a slave-owning society does result in those values, whatever their genesis, being subsumed by an encompassing idiom which continually contrasts the free and the unfree, and that such an idiom inevitably reorients those values and imbues them with a quite particular set of social connotations.

If we return to women, or rather to the 'character' of women –

to the character of those who were denied any social autonomy – does not their 'natural' susceptibility to emotions, desires, passions, and the bodily appetites mark them in Athenian thought as fundamentally unfitted to be free? Or, to put matters the right way around, is not their 'psychological' characterization made to be ideologically consistent with the place assigned to them within the social structure of the *polis*? It is instructive once more briefly to revert to Aischines' prosecution of Timarchos:

> And shall you let Timarchos go free, a man chargeable with the most shameful conduct? A man with a man's body, but having committed a woman's follies. If so, how can any of you punish a woman whom you have found to have done wrong? What man will not appear ignorant if he is angry with her who has erred in accordance with nature (*kata physin*) while he treats as an adviser [in the assembly] a man who has behaved recklessly contrary to nature (*para physin*)?
>
> (Aischines 1 [*Timarchos*]: 185)

It is all too easy here to misconstrue the meaning of Aischines' words. The references are not to 'unnatural acts' (as they were once called), but to an unnatural incontinence on the part of someone who ought to have been capable of regulating himself as a free man (cf. Dover 1978:67). And if Timarchos is to lose his civic rights because in this respect he is no better than a woman (in fact worse than a woman, since the sexes are to be judged differently), it is clear that woman's nature automatically debarred her from the exercise of political authority.

This same transposition of women's social and political condition to the realm of 'natural characteristics' can also be seen in the common charge of gluttony. Given the high degree of enforced idleness in which upper-class Athenian women lived it is possible that such a charge was not without empirical foundation. Still playing the farsighted liberal Ischomachos actually advises his wife not to sit around the house 'like a slave' but actively to supervise the household and even to join in the work in order to give herself exercise, keep herself healthy, put colour in her cheek, and give herself a good appetite (not to be confused with gluttony). But in general I would argue that the portrayal of women as gluttonous related more to a *conception* of women as naturally 'unfree' than it did to the reality of their conduct. For whatever the truth of the portrayal, the correspondence at the ideological level between this form of female self-indulgence and the stock-in-trade characteristics of the slave is striking. Both women and slaves are made to contrast with the austere ideal of the self-determined 'free' man who, in Aristophanes' *Ploutos* (190–2), dreams of love (presumably homosexual) – and his slave of bread; of literature – and his slave of sweets; of honour – and his slave of cheese-cakes; of manliness – and his slave of dried figs; of ambition – and his slave of barley-meal; of command – and his slave of pea-soup. At each point the high-minded aspirations of the free man are juxtaposed with the gustatory desires of the slave. But at each point they would also serve to contrast with the purported preoccupations of women. Women are in bondage to their physical appetites as much as those who are legally bound. The free man is contrasted in his inclinations both with those who are completely controlled (slaves) and those who are not in control of themselves (women).

Much the same argument can be made to explain women's purported fondness for drink. Ehrenberg has written, 'The countless suggestions in comedy of women's love of drink, however much exaggerated, cannot have been without some real basis; wine might be a consolation in their frequent loneliness' (1974a: 202). There is no reason to deny the possibility that women consoled themselves with wine, but again I suspect that the 'real basis' for this characterization was more than empirical. The association of sex and wine was commonplace in Greek thought, for the two were of the same order; both belonged to the unrestrained, untempered, irrational areas of life. Mythologically the centaurs in their wildness were both lewd and inebriate, and while Dionysos may have been god of wine, he was also god of inebriation in a fuller sense (nor should his dominion over women be forgotten – though I shall return to that later). And it is perhaps worth noting that in Homer the one form of *ate* not ascribed to the intervention of some god was that caused by wine (Dodds 1951: 5). At a more mundane level, however, sex and wine were the commonplace follies of irrational youth, while the coupling of *hetairai* and drink was virtually institutionalized by the *symposia*, the male drinking parties, at which the only women present were sexual entertainers. In short, I would suggest that the portrayal of women as habitual drunkards in Aristophanes' plays was appealing to something other than observation or even exaggeration; it was appealing to a conception of women as the representatives or associates of disorder, emotion, passion, irrationality. Both women's sexual avidity and their inebriation were manifestations of their innate lack of 'self-control', their enslavement to unruly desires. And again it should be noted that slaves, like women, were portrayed as secret tipplers. Indeed, as Vogt (1974: 8) summarizes, for the most part slaves were represented as lazy, insolent, randy, cowardly, thieving, and dishonest – equally the comedy attributes of women.

To suggest that this similarity between the stock vices of slaves and of women points towards some simple equation between slaves and women would, of course, be untrue. The distinction between being legally slave and being legally free was, after all, quite as radical for women as it was for men. As we have seen, the status of a free-born Athenian woman was as protected and as exclusive as that of her father, brother, or son. Nevertheless, the similarity is not without significance. However different the social position of slaves and women, and however different Athenian attitudes towards them, both could be contrasted with the male Athenian citizen, for both lacked those characteristics of self-control, restraint, indeed of moral integrity, which were the mark of a free man. Both slaves and women were motivated by the ephemeral desires of physical well-being and of immediate advantage rather than by the loftier ideals of good behaviour which involved self-denial and which were perceived through the exercise of reason. Within the broader context of freedom and slavery the similarity of their psychological portrayal is thus revealing; neither slaves nor women were held to possess by nature those qualities which were the necessary requirement of a self-governing man; and neither slaves nor women possessed the political rights that accrued to men.

In this sense, the *polis* was not only a politically exclusive entity; its members also appropriated to themselves those innate characteristics and virtues deemed necessary for those who shared the responsibilities of ruling rather than the necessity of being ruled, who had to be capable of both self-governance and self-control. And it should be noted that lack of self-control – incontinence, physical indulgence, inebriation, sensuality, luxury – are reported as the natural characteristics not only of slaves and of women but also of those *barbaroi* who lived beyond the bounds of the civilized Greek world. It is part of the complex Athenian male self-definition that barbarians are routinely characterized in their wildness and in their luxury as being *both* effeminate *and* slavish.

VIII

As usual it is Aristotle (though not himself an Athenian) who best manages explicitly to state ideas embedded in the society of his time, and on whose observations much of this chapter could be read as a mere gloss. I shall quote several passages at length, for Aristotle's analysis of 'political' life brings into focus precisely the connection that I have been trying to suggest between the position of women and the position of slaves. The difference between my comments and Aristotle's lies only in the level at which they are situated. Aristotle purports to be explaining facts of nature and their social consequences; I have been attempting to describe an ideology (of which Aristotle's explanations are themselves a part).

In his discussion of household management (*Politics*: 1259a-b), Aristotle isolates three types of 'rule': of master over slave, of father over children, of husband over wife. That of master over slave is 'despotic'; that of father over children is 'royal'; that of husband over wife is 'political'. As a form of political rule, however, that of husband over wife is somewhat exceptional in that there is never an interchange between the role of ruler and of ruled: 'As between male and female this relationship of superior and inferior is permanent.' Nevertheless, in distinguishing between 'despotic', 'royal' and 'political' Aristotle is at pains to show that the authority of a man over his slaves, over his children, and over his wife is in each case based on a different principle, and this principle in turn relates to the nature of their respective *psyches*:

> *In the psyche the difference between ruler and ruled is that between the rational and the non-rational.* It is therefore clear that in other connections also there will be natural differences. And so generally in cases of ruler and ruled; the difference will be natural but they need not be the same. For the rule of free over slave, male over female, man over boy, are all natural but they are all also different, because, while parts of the *psyche* are present in each case, the distribution is different. *Thus the*

> *deliberative faculty of the psyche is not present at all in the slave; in a female it is inoperative* (akyron), *in a child, undeveloped.*
>
> (Aristotle, *Politics*: 1260a; T. A. Sinclair trans., modified; my emphasis)

Aristotle thus continues to stress that slaves, women, and children, although all naturally inferior, and although all subject to rule, cannot be treated as a single group. Yet at this 'psychological' level the argument seems curiously weak. In distinguishing children (by which he means *male* children) from all the rest he is on safe ground, for as he has already pointed out (1259b) the difference between a (male) child and his father is temporary whereas the differences between slaves and their rulers and women and their rulers are permanent.

But in terms of his own argument the further separation between slaves and women is not so easy to maintain. Both fall permanently on one side of what, for Aristotle, is a fundamental division: the division between those who are 'rational' and those who are 'non-rational'. As he has stated in an earlier passage:

> The rule of mind over body is absolute, the rule of intelligence over desire is constitutional and royal. In all these it is clear that it is both natural and expedient for the body to be ruled by the mind, and for the emotional part of our nature to be ruled by that part which possesses reason, our intelligence. The reverse, or even parity, would be fatal all round. This is also true as between man and other animals; for tame animals are by nature better than wild, and it is better for them to be ruled by men; for one thing, it secures their safety. Again, as between male and female, the former is by nature superior and ruler, the latter inferior and subject. And this must hold good of mankind in general. We may therefore say that wherever there is the same wide discrepancy between two sets of human beings as there is between mind and body or between man and beast, then the inferior of the two sets, those whose condition is such that their function is the use of their bodies and nothing better can be expected of them, those I say are slaves by nature. It is better for them, just as in the analogous cases mentioned, to be thus ruled and subject.
>
> (ibid., 1254b)

It is quite clear from this passage that the relationship between man and woman is of the same order as that between mind and body or between man and beast, and that consequently as a 'set of human beings' women are 'slaves by nature'. They are permanently on the non-rational side of the rational/non-rational dichotomy – precisely the division between rulers and ruled. One might therefore be led to think that the distinction Aristotle goes on to make between the *psyche* of slaves and the *psyche* of women on the grounds that in the former the rational faculty is 'absent' while in the latter it is 'inoperative' or 'without command' (*akyron*) is something of an *ad hoc* refinement to his theory which stems from a reluctance to follow where his own logic leads. More kindly one might say that the empirical side of Aristotle's philosophy takes over and that he is anxious to account for the fact that in the society in which he lived it was manifestly the case that women and slaves were not placed on the same footing either in terms of their social rights or in terms of the respect which they commanded. Indeed, Aristotle makes this very point, but in a most intriguing manner:

> Nature has distinguished between female and slave. She recognizes different functions and lavishly provides different tools, not an all-purpose tool like the Delphic knife; every instrument will perform its work best when it is made to serve not many purposes but one. So it is with the different functions of females and slaves. Some non-Greek communities fail to understand this and assign to female and slave exactly the same status. This is because they have no section of the community which is by nature fitted to rule and command; their society consists solely of slaves, male and female. So, as the poets say, 'It is proper that Hellenes should rule over barbarians', meaning that barbarian and slave are by nature identical.
>
> (ibid., 1252a34–69)

It is thus only in civilized communities – that is to say, *Hellenic* communities – that the distinction between slaves and women is properly apparent; and it is apparent because in such communities there is a section of the community who are by nature free and fit to rule and who can perceive and impose the proper relationships between categories of people. Aristotle's argument here is curiously elliptical. The two meanings of 'slave' are compressed so that in barbarian communities women are reduced (wrongly) to the same status as slaves while in such communities *everybody*, males and females (and presumably both 'free' and 'slave') are without differentiation slavish. But if the argument is elliptical it is also richly suggestive, and it would not be unwarranted to say that for Aristotle among the undifferentiated mass of the barbarians not only was there no distinction between the characteristics of free men and of slaves, but also none between those of men and of women. The 'natural' characteristics of slaves, women, and barbarians all merge in their opposition to the qualities of the free men of the civilized Greek *polis* – the citizen body of men in the true sense: controlling and self-controlled.[17]

IX

Let me briefly recapitulate. I have argued that three pairs of interrelated oppositions were fundamental to Athenian moral thought: freedom/slavery, self-control/incontinence, rationality/emotion. Women were consistently portrayed as undisciplined and emotional; they were therefore by nature 'unfree' – that is to say, they lacked the necessary qualities for self determination and autonomy. In the extended sense of the word they were 'slavish' – not as members of the legally defined category of slaves but in terms of their innate characteristics. Nevertheless such a characterization was not without its social concomitants. The very use of 'slavishness' as a general classificatory device stemmed from the actual existence of a class of people who were legally slaves and who possessed no rights. And, if in terms of their psychology, women were thought innately to possess slavish characteristics, in terms of their social and legal position within the *polis* they also possessed few *independent* rights. To use Aristotle's terminology, if the rational faculty of women's *psyche* was *akyron*, 'inoperative', 'without command', then the obvious solution to the problem of their very necessary accommodation within the organization of a civilized community was to place them under a *kyrios*, a male, who could supply for them that rational command which they lacked. This is precisely the situation envisaged by Athenian law and social organization whose provisions have been examined.

It must be stressed, however, that no direct causal relationship can be posited in either direction between women's position within the organization of the Athenian *polis* and the Athenians' conception of women's natural characteristics. The most that can safely be said is that the social and legal provisions pertaining to women and the commonplace characterization of female nature were consonant. If a fairly close relationship can be seen to exist between Athenian ideas about the nature of women and the social and legal provisions which regulated their conduct, the cohesion is best understood in terms of a mutual accommodation whereby existing social relationships could, if necessary, be justified by an appeal to the 'innate' characteristics of the female species, while social and legal provisions themselves were able to provide evidence about the 'natural' characteristics of those whom they embraced. If it can be said that the problem of integrating women – the rational faculty of whose *psyche* was *akyron* – into the organization of a civilized community was solved by placing them under the supervision of men to act as their *kyrioi*, this is not to suggest that the problem ever presented itself to the Athenians in quite this manner and, having been thus presented, was summarily resolved. Both the problem and its solution constitute, so to speak, a *post factum* re-presentation of existing states of affairs – a body of ideas about the nature of women, a body of social legislation and custom – which had perhaps largely autonomous histories but which were drawn into a mutually sustaining and mutually revealing combination. Indeed, the connection between Aristotle's use of the term *akyron* in describing the state of women's rational faculty and the existence of guardians called *kyrioi* under whom women were placed is perhaps no more than linguistically fortuitous, but it does at least serve neatly to underline the manner in which Athenian ideas about the nature of women and the social and legal institutions which regulated women's conduct can be seen to form a unity. That unity should be stressed.

As we have seen, women's membership of the Athenian *polis* was always derivative, dependent on their associations with the men through whom they gained their status and their rights. Like other non-citizen groups of the Athenian population their presence was necessary for the existence of that state – vitally so, since they bore its progeny and transmitted political rights among them – but they were not in their own right members of the *polis* which remained '*un club d'hommes*'. They were members of the *oikoi* of those who were members of the state. Their position was, as it were, marginal. They existed on the peripheries of political life – necessary for its maintenance, excluded from its activities. At the level of psychological representation women's position was homologous. They were endowed with those characteristics of sensuality, irrationality, emotionality which, though recognized as always present in and even necessary to human existence, had to be restrained, controlled, and subjugated if civilized life was to be maintained. Women's natural characteristics were those which were peripheral to the character of free, self-governing, and autonomous men. Just as women were in an important sense outside the *polis*, so the characteristics which they possessed were in an equally important sense outside the nature of its *politai*.

9

THE ENEMY WITHIN

I

Euripides has been variously judged a misogynist and the champion of women – a division which, at the least, testifies to the complexity of his female portrayals. What is certain is that in the plays of Euripides (perhaps more clearly than in the works of any other Greek author) women appear as the perpetrators of crimes motivated by passion, jealousy, and vindictiveness, and committed with deceit and guile. The complicating factor is that Euripides also presents women's actions as emanating from the insecurity of their position. Vindictive ferocity stems from a social impotence in which deception and guile are the only weapons and where jealousies and hatreds have been born of justified fear.[1] The *Ion* is a case in point.

As we learn from the god Hermes' opening speech, Kreousa, daughter of King Erechtheus of Athens, was raped by the god Apollo to whom she bore a son. To protect her reputation she exposed the infant at birth. Later, she was married to Xouthos, a foreigner, who became ruler of Athens. Together they have led a life of childless sorrow. Kreousa's story has thus started with sexual injustice – Apollo's rape and her cast-out child – and has continued with injustice, for she must remain childless while retaining her secret guilt. As Euripides has her say on entering the stage:

> O the wretchedness of women. O the recklessness of gods. Well, where shall we go for justice if we are crushed by those who are our masters?
>
> (Euripides, *Ion*: 252–4)

Moreover, for women, as Kreousa knows, the injustice of the gods is matched by that of men:

> For the lot of women is troublesome to men and since the good are confused with the bad we are hated. Thus we are born unlucky.
>
> (ibid., 398–400)

Such masculine injustice, both human and divine, is compounded as the play's action begins, for Xouthos and Kreousa have come to Delphi to entreat the god Apollo to end their childlessness. The oracle tells Xouthos that the first man he sees will be his son; that man, of course, is Ion, in fact Kreousa's own child by Apollo, who was saved and brought up as the god's servant in his sanctuary. The oracle, however, makes no mention of the circumstances of Ion's birth and Xouthos decides that Ion must be his own illegitimate son, the progeny of some youthful escapade. Overwhelmed with joy he wishes to install Ion as his heir and as the future ruler of Athens. But Ion has reservations, and not least of these is the hatred he feels sure he must incur from Kreousa when she becomes his 'stepmother':

> And coming to a strange house as a foreigner and to a childless woman who previously has shared her sorrows with you, but will have to bear her lot bitterly by herself, how shall I not with reason be hated by her when I stand close by you at your foot and she, childless, bitterly looks on at your affection?
>
> Either you must cast me off and look to your lady, or honour me and throw your household into confusion.
>
> How often have women found daggers and bowls of deadly poison for their men.
>
> (ibid., 607–17)

Ion's view on the likely conduct of women when they feel themselves displaced or threatened is clear: they are potential homicides. At the same time, however, the speech is both compassionate and understanding. How could Kreousa not loathe and wish to destroy the person whose presence destroys her own fragile world? But the compassion does not alter the facts, and Ion's prediction of Kreousa's reaction proves only too correct. Kreousa is outraged at the thought of Xouthos' bastard son being introduced into their house; equally, she is terrified of being estranged from her husband's affections and discarded. She does indeed plot to poison Ion – but again at this point the chorus of justification that her female servants sing should be noted:

> Mark this you who always chant
> with your shrill songs
> about our beds, our unrighteous
> unholy unions of love –
> how much do we surpass in righteousness
> the unjust ploughings of men.
> Let discordant song
> and the shrill
> muse attack men and their beds
> (Euripides, *Ion*: 1090–8)

The theme of exploitation is returned to. First Kreousa was the victim of Apollo's lust; should she now pay the price for Xouthos' philanderings?

In the *Ion*, as in other plays, Euripides seems to be querying the values of a society in which the interests of women are overridden so that they are constrained to defend themselves by stealth and crime, while simultaneously reaffirming their readiness to do so. When the news of Ion's discovery is made known, Kreousa's old slave offers this advice:

> Now you must do something truly female:
> take a sword, or by some guile
> or poison slay your husband and his son
> before death comes to you from them.
> (ibid., 843–6)

The slave reiterates precisely Ion's view of what is 'truly female', and if perhaps we are made to feel that the phrase here amounts to no more than an incidental cliché, nevertheless it is on her slave's advice that Kreousa acts.[2] Euripides may make us sympathetically aware of the condition of women so that we see their impassioned crimes as the desperate ploys of those bent on preserving their rights in the one area of life where they hold some sway, but the unreasoning violence of their emotions is still maintained. Very much like Dover's characterization of the poor Athenian male, Euripides' women are 'abject and desperate by turns, afraid to take a chance yet compelled to do so, and at the same time resentful and vindictive'. As such, women, like all beings incapable of exercising freedom, still form the 'antithesis of the self-sufficient, self-reliant character which was admired, the character of a *man* who can choose and afford to give' (1974: 110; my emphasis).

It is perhaps the tragedy of Euripides' women that they are continually forced to live out the very role in which male prejudices cast them. Kreousa does become the murderous stepmother to which stereotype Ion refers (just as in the *Hippolytos* Phaidra becomes the love-crazed female whom Hippolytos so despises). We are given a series of relatively complex and sympathetic insights into the motivations of female actions, but the objective results of those actions serve only to confirm the misogynist clichés with which Euripides' plays are spattered: and these would have us believe that in the passionate nature of women there lurks the constant threat of violence and injury to men. Moreover, that threat of violence and injury derives not just from their individual predicaments or from their individual characters, but, as the clichés remind us, is inherent in their nature as women – a nature which is not rational, which is not self-controlled, which is not 'free'.

Euripides (like all the Greek dramatists) presents situations which are extreme, which, if measured against the day-to-day reality of Athenian life, are fantastical. They nevertheless contain an imaginative construction of what women would be like if ever (or whenever) they were to slip from the control of a society which, on the one hand, has constant need of them but which, on the other, must repress the irrationality, the emotion, the passion, the disorder which they introduce into its midst. That view, of course, is not Euripides' alone. Let me return to a relatively mundane situation.

II

I have already commented on the passage from Demosthenes 54 [*Konon*]:

> when my bearers got to the door, my mother and the women slaves began shrieking and wailing, and it was with difficulty that at length I was carried to a bath.
>
> (Demosthenes 54 [*Konon*]: 9; Loeb trans., modified)

The speaker is not criticizing his womenfolk; nevertheless he implies that their behaviour was less than helpful. And if, as we have seen, the 'natural' inclination of women to wail and grieve in their distress could be institutionalized by Athenian society, it could also be seen as a threat to the order of that society, for an interesting law ascribed to Solon actually limited the participation of women in funerals:

> The deceased shall be laid out in the house in any way one chooses, and they shall carry out the deceased on the day after that on which they laid him out before the sun rises. And the men shall walk in front when they carry him out, and the women behind. And no woman less than sixty years of age shall be permitted to enter the chamber of the deceased, or to follow the deceased when he is carried to the tomb, except those who are within the degree of children of cousins; nor shall any woman be permitted to enter the chamber of the deceased when the body is carried out, except those who are within the degree of children of cousins.
>
> (Demosthenes 43 [*Makartatos*]: 62; Loeb trans., modified)

The exact purpose of Solon's law must remain in doubt, but following Alexiou (1974: 21–2) and Humphreys (1983: 85–6) I think it unlikely that it formed part of any deliberately 'anti-female' programme of legislation. More likely the motivation was political, part of Solon's attempt to limit the influence of aristocratic families for whom lavish funeral processions would have provided the opportunity to display the extent of their following. Solon was trying to ensure that funerals remained discreetly private affairs. Nevertheless, the form that his legislation takes is revealing, for if the above interpretation is correct, it constitutes a clear recognition that what could transform a private funeral into an emotionally fraught and hence potentially dangerous public demonstration was precisely the uninhibited conduct of women. If the charged atmosphere of such processions was to be defused, then it was the participation of women which had to be limited.

The disruption caused by women's lamentations, the threat they pose to the good order – indeed to the very survival – of society by their introduction of the unrestrained, the emotional, the illogical is again imaginatively illustrated in extreme form by Eteokles' outburst in Aischylos' *Seven Against Thebes*. As the army of Polyneikes is heard to approach the city's gates, the chorus of Theban women presents a song of supplication and terror to the ancestral gods – a song which Eteokles tries to silence:

> Insufferable creatures, I ask you if this is what's best for the safety of the city, if it gives courage to our beleaguered army – to fling yourselves before the images of our protecting gods, shouting, shrieking, making yourselves detestable to those who are right minded/constrained (*sophrones*). Neither in troubles nor in happy prosperity would I dwell with the race of women. If woman has the upper hand, her daring (*thrasos*) cannot be tolerated; but if she is afraid she is an even greater evil to household and city. Now with your shouting and running everywhere in flight you have set mindless cowardice (*apsychos kake*) amongst the citizens. Thus you aid the very enterprise of those who are without, and we are ruined from within by our own selves.
>
> (Aischylos, *Seven Against Thebes*: 181–95)

The confrontation between Eteokles and the women continues:

> *Women*: I am afraid. The battering grows louder at the gates.
> *Eteokles*: Will you keep quiet and say nothing of this in city.
> *W*: O company of gods, abandon not our battlements.
> *E*: A curse on you, can you not suffer this in silence?
> *W*: Gods of the city, let me not be enslaved.
> *E*: It is you who are enslaving me and the whole city.
> *W*: Almighty Zeus, turn your bolt on the enemy.
> *E*: O Zeus, what a race of women you have given us.
> *W*: A miserable race – like men whose city has been taken.
> *E*: Will you speak ill-omens with your hands on the statues of the gods?
> *W*: With faintness of mind (*apsychia*) fear takes hold of my tongue.
>
> (ibid., 249–59)

Eteokles' tirade repeats the commonplace accusations: women are naturally fearful, naturally given to panic, always emotional, lacking constraint, and to be contrasted with those who are of sound mind (*sophrones*). The panic they are spreading is 'mindless' (*apsychos*); the women themselves admit that they have been overcome by *apsychia*. But it is difficult not to read the exchange at a more abstract level. When the women express their fears of falling into slavery, Eteokles' retort is that it is the women themselves who are enslaving the city by their behaviour. And though the assimilation of the women to Thebes' external enemies has an immediate contextual relevance, one wonders whether it is not intended to be more permanent and complete. Eteokles makes it clear that the same characteristics he abhors in the present circumstances are manifested by women even in peacetime. Are not women always outside the boundaries of the *polis* by virtue of the emotionalism and irrationality they represent? Do they not constitute within society and as a group (a 'race') the very analogue of those irrational forces which seem to lie outside the rational psyche and which threaten to enslave it? In the immediate circumstances the survival of the *polis* depends on keeping the destructive forces of the enemy outside Thebes' walls; but at a conceptual level the survival of the *polis* depends on keeping the destructive forces of emotionalism and irrationality outside the city's definition. If Eteokles can claim that the women are enslaving the city, is this because their behaviour will result in the conquest of Thebes by Polyneikes' army, or is it because their behaviour already amounts to a manifestation of Thebes' submission to the slavery of the emotions, to a renunciation of self-control?

Aischylos' dramatization will not allow simple judgements to be formed. If Eteokles gives voice to a view of women which makes them the natural enemy of civilization and of order, there is equally little doubt that Eteokles' own behaviour reveals a dangerous lack of restraint. In contrast to Eteokles' pronouncements, Aischylos seems to assure us that women do have a legitimate place in society. Moreover, another paradigm of the female role plays counter-point in the play to Eteokles' uncompromising view. The contrast between 'those within' and 'those without', between the besieged inhabitants of Thebes awaiting the outcome of battle and the reported activities of the enemy forces drawn up against the city's gates, is fundamental to the dramatic structure of the play. In this connection the fact that Aischylos chooses a chorus of women is not without importance. The women represent not simply one side of a military conflict: the beseiged Thebans as opposed to Polyneikes and his Argive army. They represent one side of life: the settled inner and domestic world of the enclosed *polis* as opposed to the external chaos of heroic ambition and war. As Eteokles exits, the chorus sings of the fate of the fallen city, and it is through the voices of women that Aischylos can convey the full meaning of that event: the total destruction of a community.

Here, then, there is something of a reversal. Whereas with Eteokles the women are represented as in allegiance with the destructive forces that lie outside the boundaries of the *polis*, now it seems that the women represent the very values of the *polis* in danger of being destroyed. If this is a paradox, then it is one of the paradoxes on which the play is built. But it should be noted that the reversal relates not only to the position of women; Eteokles too becomes an ambivalent figure. His hostility towards the women of Thebes places him in opposition to the city's values of kinship and religion which they represent. The women pray for safety and for the salvation of their city:

> But it was with trust in the gods that I came in hot haste to the olden images of the divine powers when there arose the roar of the deadly hail beating on the gates. Then indeed I was stirred by terror to supplicate the blessed ones that they might hold protection over the city.
>
> (ibid., 211–15; Loeb trans., modified)

But Eteokles argues that:

> It is for men to offer victims and sacrifice to the gods when they are encountering the foe; your task is to keep quiet and stay within the house.
>
> (ibid., 230–2)

And although he reluctantly allows women's right to supplicate the city's gods, he adds that:

> If then you hear of dying or of wounded men, do not seize the news with shrill lament. This is what Ares feeds on – the blood of mortals.
>
> (ibid., 242–4)

Eteokles wants success, not safety, and the bloody world of action is his chosen domain.

Thus while Eteokles accuses the women of being destructive of discipline and order within the city, it is his own hatred that shows the greater lack of self-restraint. And if he lays the possibility of Thebes' defeat on the women's heads, it is the world of masculine action and ambition that will be the cause of destruction. Going personally to defend Thebes' seventh gate, he pits himself against Polyneikes, his own brother, as if determined to fulfil the curse of his father Oidipos. And now it is again the chorus of women who see the folly of his actions and who preach moderation; who tell him not to let the madness (*ate*) of battle carry him away (686–8); who tell him that too savage a passion is impelling him to slaughter of unlawful blood (692–4); who order him not to defend the seventh gate himself (714); who, in short, tell him to 'be ruled by a woman though you like it not' (712). But for Eteokles, 'No soldier can brook utterances like these' (717). As for the city's gods:

> The gods it seems have long since ceased their care of us. The service they value at our hands is that we perish. Why then should we any longer fawn upon the fate of death?
>
> (ibid., 702–4)

The Theban women's lack of self-control endangers the state; but if the women represent a disorder that Eteokles as ruler of Thebes must repress, it is nevertheless clear that women are also the representatives of an order that Eteokles fails to grasp. Eteokles asserts his own ambitions, his own strength, his own will, his own destiny, perhaps even his own individuality; the women fall back on the solidarity of kinship and of the identity of a city with its gods. There is a tension between what is an essentially masculine and 'political' order and an essentially feminine and 'natural' one; and it is a tension that continues throughout the play.

Thebes is saved, but Eteokles and his brother Polyneikes kill each other. The chorus of women once more breaks into lamentation:

> Our loud-resounding, piercing wail attends them – a wail of unforced sorrow, unforced pain, from minds distressed, wherein there is no thought of joy, and poured forth with tears in all unfeignedness from out our heart that wasteth as I weep for our two princes here.
>
> (ibid., 915–21)

At this point Eteokles and Polyneikes' two sisters, Antigone and Ismene, appear on stage to sing a dirge over the corpses of their brothers (957–1010). These lamentations over, a herald enters proclaiming a decree of the Council of Thebes: Eteokles is to be buried with honour on Theban soil; Polyneikes, a traitor, is to be cast unburied from the city, a prey to dogs. And again the conflict between men and women, between the two orders of the community, erupts. Antigone's words speak for themselves:

> But I declare to the rulers of the Kadmeans – if none other be willing to take part with me in burying him, I will bury him, and I will risk the peril of burying my own brother; nor do I feel shame thus to be an unsubmissive rebel to the State. Strange power – the bond of common blood whence we are sprung, from wretched mother and unhappy father. Therefore, my soul, in sisterly loyalty of heart, take willing part with him in his distress who has no will, the living with the dead. His flesh no gaunt-bellied wolves shall tear – let no one 'decree' me that! Even if I am a woman, I will find him a burial and a grave, bearing the earth in the folds of my linen raiment. With my own hands I will cover him. Let no man 'decree' it otherwise. Courage! I shall find the means to act.
>
> (ibid., 1032–47)

It is arguable that this last scene of the *Seven Against Thebes* is spurious, a later addition to Aischylos' play, modelled on Sophokles' *Antigone*,[3] but whether the last scene is integral to Aischylos' play or retrospectively derived from Sophokles', the work illustrates two competing and to an extent contradictory paradigms of the female role that find expression in Athenian drama. The one sees woman as a threat to the order of society by virtue of her uncontrolled nature; the other sees her as the guardian of those values of kinship and religion which are integral to the order of society. Both, however, are rooted in the structure and thought of the Athenian *polis*, and drama seems to have seized now on the one, now on the other in its imaginative constructs.

The supposedly passionate, sensual, unrestrained, irrational nature of women, their exclusion from the political life of the state, their confinement to a purely domestic role as the irresponsible subordinates of their male *kyrioi*, could all be translated by imaginative literature into a vision of women as either the deliberate or the accidental allies of those forces which, either physical or, as it were, spiritual, lay outside the boundaries of the civilized *polis*. At the same time, the very nature of women's incorporation into society – not as autonomous and independent individuals but as permanent members of the secluded familial world – could also make them seem the representatives of values more enduring and perhaps more 'natural' than those achieved by men in their struggle to rule themselves and others. Even without its final scene the *Seven Against Thebes* seems to interplay both conceptions. It is, however, in Sophokles' *Antigone* that they are placed in sharpest juxtaposition.

In Sophokles' play, a dramatic sequel to the *Seven Against Thebes*, Kreon, tyrant of Thebes, has ruled that Polyneikes is to remain unburied as punishment for his treachery. Antigone, Polyneikes' sister, buries him and is thus condemned to death. The conflict between traditional religious obligations towards family and kin, and the necessity of obedience to the legitimate authority of the state forms the substance of the play. The nature of the conflict is clearly expressed by Antigone:

> For me it was not Zeus who made that order.
> Nor did that Justice who lives with the gods below
> mark out such laws to hold among mankind.
> Nor did I think your order were so strong
> that you, a mortal man, could over-run
> the gods' unwritten and unfailing laws.
> Not now, nor yesterday's, they always live,
> and no one knows their origin in time.
> So not through fear of any man's proud spirit
> would I be likely to neglect these laws,
> draw on myself the gods' sure punishment.
> I knew that I must die; how could I not?
> even without your warning. If I die
> before my time, I say it is a gain.
> Who lives in sorrows many as are mine
> how shall he not be glad to gain his death?
> And so, for me to meet this fate, no grief.
> But if I left that corpse, my mother's son,
> dead and unburied I'd have cause to grieve
> as now I grieve not.
> And if you think my acts are foolishness
> the foolishness may be in a fool's eyes.
>
> (Sophokles, *Antigone*: 450–70; E. Wyckoff trans.)

Equally, however, the conflict is presented as one between women and men.[4] As Kreon says to Antigone:

> Then go down there, if you must love, and love the dead. No
> woman rules me while I live.
>
> (ibid., 524–5)

And as he later explains to his son, Haimon:

> There is no greater wrong than disobedience.
> This ruins cities. This tears down our homes,
> this breaks the battle-front in panic-rout.
> If men live decently it is because
> discipline saves their very lives for them.
> So I must guard the men who yield to order,
> not let myself be beaten by a woman.
> Better, if it must happen, that a man
> should overset me.
> I won't be called weaker than womankind.
>
> (ibid., 672–80)

The play leaves us in little doubt that in this conflict between the arbitrary 'political' laws of men and the unwritten but axiomatic laws of the gods, Kreon is in the wrong – and suffers for it (cf. Lefkowitz 1986: 81). His son, Haimon, suicides; his wife, distraught at her son's death, suicides in turn, cursing her husband. As Pomeroy has written:

> *Antigone* and many other tragedies show the effect of the over-evaluation of the so-called masculine qualities (control, subjugation, culture, excessive cerebration) at the expense of the so-called feminine aspects of life (instinct, love, family ties) which destroy men like Kreon. The ideal, we can only assume – since Sophokles formulates no solution – was a harmonization of masculine and feminine values, with the former controlling the latter
>
> (1975: 103).

But it is the lack of a solution which should be stressed, for while Antigone, the woman, represents and upholds the conservative values of kinship and religion, and while Kreon, the man, suffers for his actions, there is still no doubt that the love, the passion, the emotionalism which such women as Antigone introduce into the affairs of men remain destructive forces. As Pomeroy again notes

(1975: 102), the fate of Haimon, who dies for love of Antigone, underscores the fact. The play's chorus comments:

> Love unconquered in fight, love who fall on our havings.
> You rest in the bloom of a girl's unwithered face.
> You cross the sea, you are known in the wildest lairs.
> Not the immortal gods can fly,
> nor men of a day. Who has you within him is mad.
> You twist the minds of the just. Wrong they pursue and
> are ruined.
> You made this quarrel of kindred before us now.
> Desire looks clear from the eyes of a lovely bride.
> (Sophokles, *Antigone*: 781–96; E. Wyckoff trans.)

Thus, side by side with the portrayal of women as the righteous defenders of the axiomatic values of kinship and religion there is still the portrayal of women as the natural enemies of order and the instigators of strife. If Antigone defends the laws of kinship and religion, Antigone also introduces into the life of Kreon, his city, and his family that passion which enslaves and kills his son and destroys his wife. The defender of kinship is the ruination of Kreon's family, and women still represent within society those dangerous emotions better left outside it.[5] Kreon's edicts may have been misguided, but that the challenge to them comes from a woman is based, I think, as much on a general conception of women as disruptive of civic order as on the fact that they were linked to the values of family and kin. Moreover, what is so often put in question by Greek tragedy is not, I suspect, the validity of male authority and the attempt rationally to regulate life but the limits of that enterprise's success in a fundamentally irrational world. In the *Antigone* the failures of male order and control are revealed by a woman because it was always the irrational which tested the civilization of men. Women lack self-control and so must be controlled; every time that control is broken (even justifiably) havoc ensues.

Literary works are seldom, of course, political or social manifestos. It is the complexity of their portrayals, not the clarity of their pronouncements, which earns them their place. Sophokles' *Antigone* is not simply about males and females; indeed it is not 'simply' about anything. I have chosen in the above analysis to underscore a particular relationship, that between masculine and feminine, that the play exploits. This is by no means to assert that this relationship is what the play is 'about'; nor should my argument be taken to suggest that Sophokles' play amounts to a defence of masculine control. Like most literary works the *Antigone* manipulates the cultural assumptions of its time. It is those cultural assumptions with regard to women that I am concerned to stress. The play itself both uses them and goes beyond them.[6]

III

Aristotle's *Politics* takes up much the same point, though in a purportedly historical rather than literary context. For, as a student of society, he demonstrates to his own satisfaction how another aspect of women's lack of self-control could, if unchecked, lead straight to social decline. In Athens the adornment of women was probably one of the more available forms of conspicuous consumption among the upper classes, since men prided themselves on the simplicity of their dress. But having transferred wealth's ostentation to women, Athenian society then standardly reproached them for their extravagance in clothing and jewellery (Dover 1974: 179–80). When Aristotle turns to a critique of the Spartan constitution, it is again in women's nature (and men's negligence) that he sees an outstanding flaw:

> For there the law-giver, whose intention it was that the whole of the state should be firm in character, has certainly taken some pains as far as the men are concerned, but he has failed and been negligent over women. For at Sparta women live without restraint, enjoying every licence and indulging in every luxury. One inevitable result of such a way of life is that great importance is attached to being rich, particularly in communities where the men are dominated by the women.
>
> (Aristotle, *Politics*: 1269b; T.A. Sinclair, trans. modified)

It is possible that genuine differences existed between the economic and social position of women in Sparta and women in Athens,[7] but the interesting feature of Aristotle's treatment of the subject is his assumption that if women are not carefully regulated then they will automatically indulge in every form of 'licence and luxury'. Aristotle offers a particular historical explanation for the lamentable state of affairs in Sparta:

> Hence the condition of affairs at Sparta, where in the days of

> their supremacy a great deal was managed by women. And what is the difference between women ruling and rulers ruled by women? The result is the same. Boldness is not a quality useful in the affairs of daily life, but only, if at all, in war. Yet even here, the influence of Spartan women has been harmful. This was demonstrated when Lakonia was invaded by the Thebans; instead of playing a useful part, like women in other states, they caused more confusion than the enemy. Now it is not surprising that from the earliest times lack of control of women was a feature of Lakonian society. There were long periods when Spartan men were obliged to be absent from their own land on military service, fighting against Argives, or against the Arkardians or Messenians. When they returned to a life of ease, predisposed to obedience by military life, which offers scope for many kinds of virtue, they readily submitted to the law-giver, Lykourgos. But not so the women. It is said that Lykourgos endeavoured to bring them under his control, but when they resisted, he gave up the attempt.
>
> (ibid., 1269b–70a)

But Aristotle also concludes with a general reminder about the principles of political life and the proper place of women:

> as has been said earlier, the position of women at Sparta is wrong. It offends against the principle of polity itself, and it contributes largely to the existing greed for money.
>
> (ibid., 1270a)

and it should be noted throughout how reminiscent his observations are of Eteokles' views in the *Seven Against Thebes*. If women gain the upper hand, then the inevitable result is a perversion of the values of society and a weakening of its constitution. Moreover, the specific historical circumstances which Aristotle relates serve only to explain how Spartan women were in a position to indulge their 'natural' tendencies, not how they came to possess them. In fact women are characterized by 'boldness' or 'rashness' (*thrasytes*), not a quality useful in daily life – and certainly not one to be confused with *andreia*, 'bravery/manliness'. Such boldness is merely thoughtless daring which finds its counterpart in confusion and panic.

Thus even in war-time Spartan women, who have been left to themselves, who have escaped the proper and necessary authority of men and of the law-giver, cause by their lack of discipline 'more confusion than the enemy'.[8] As Eteokles cried, in times of peace women's daring or insubordination (*thrasos*) is intolerable; in times of war their fearfulness makes them even worse. Indeed, without the complexities and qualifications to be found in dramatic representations, Aristotle's observations on the women of Sparta also recall what Kreon encountered in the *Antigone* – women grown too bold to be suppressed. Spartan men, leading an organized life in military service, are predisposed to obey Lykourgos' laws; the women, too long allowed a freedom in which, as females, they indulge in 'every licence and luxury', prove totally intractable. But Aristotle makes one further point – indeed, it is the main point of his brief argument: the Spartan women's liberty has largely contributed to their society's 'existing greed for money'. Women's boldness and women's panic are neatly fused by Aristotle with avarice, for all are signs of incontinence, of a feminine lack of self-control, and all have led to the decline of the Spartan state.

IV

For one last example of the threat which women were seen to pose to the order of society by virtue of their irrational, passionate, and uncontrolled natures, I shall turn not to imaginative literature or the reflections of a philosopher, but to a firm provision of Athenian law. The law is stated in the speech Demosthenes 46 [*Stephanos II*]:

> Any citizen . . . shall have the right to dispose of his property however he wishes if he has no legitimate male offspring, unless he is not of sound mind as a result of one of these things: madness, old age, drugs, disease, the influence of a woman, or unless he is constrained by bonds
>
> (Demosthenes 46 [*Stephanos II*]: 14; Loeb trans., modified)

but its workings are apparent in other examples of law-court oratory.

From the law itself it is again apparent how closely freedom – the exercise of civic and political rights – and rationality were allied. Such external constraints as 'necessity' and imprisonment are grouped together with madness, senility, drugs, disease, and the 'influence of a woman' as grounds for the invalidation of a will. The former socially or physically deprived a man of the ability to act on his own behalf; the latter affected his reason with the same result. But the immediate interest of this law lies in the provision that no man 'under the influence of a woman' could make a legal disposition of his property, and here the manner in which women were thought to subvert the proper functioning of society is clearly evident, for a woman's influence is specifically classed with those other conditions – madness, senility, drugs, disease – which rendered a man *paranoon*, 'not of sound mind'. The influence of a woman – like the emotions and passions to which she was herself so subject – affected the rational self. Women were the representatives and the cause of those madnesses (*maniai*) which made men no longer masters of themselves; and since with respect to the transmission of property and the rights of legitimate inheritance what affected the family affected the state, the law saw fit to invalidate the testamentary acts of men who had become a woman's prey. As the speaker of Isaios 6 [*Philoktemon*] complains:

> the woman who destroyed Euktemon's reason and laid hold of so much property is so insolent that . . . she shows her contempt not only for the members of Euktemon's family, but also for the whole *polis*.
>
> (Isaios 6 [*Philoktemon*]: 48; Loeb trans., modified)

In fact the destruction of a man's reason – and consequently of his property – through infatuation for some woman appears as a common law-court tale. Thus in Isaios 3 [*Pyrrhos*] the speaker argues:

> Let us next consider the circumstances in which it might be conceived that a marriage with such a woman might have taken place, supposing that such a thing really did happen to our uncle; for young men before now having fallen in love with such women and being unable to control themselves have been induced by folly to ruin themselves in this way.
>
> (Isaios 3 [*Pyrrhos*]: 17)

while in Demosthenes 48 [*Olympiodoros*] the following plea is made:

> But he is deranged, men of the jury, and out of his senses. . . . For you must know, men of the jury, that Olympiodoros has never married an Athenian woman according to your laws; he has no children nor has he ever had any. But he keeps an *hetaira*

> whose freedom he has bought, and it is she who is the ruin of us all and who drives the man on to a higher pitch of madness.
>
> (Demosthenes 48 [*Olympiodoros*]: 52–3; Loeb trans., modified)

Olympiodoros is mad in the speaker's opinion because he is ruled by a *hetaira* and his madness is demonstrated by a neglect of his kinsmen's interests; but in the continuation of the passage, which details Olympiodoros' neglect, that specific consequence of rule by women which Aristotle so condemned becomes apparent:

> I am striving not in my own interests only, but also in the interests of her to whom I am married, his [Olympiodoros'] own sister, born of the same father and the same mother, and in the interest of his niece, my daughter. For they are wronged not less than I, but even more. Can anyone say they are not wronged and are suffering outrageous treatment, when they see this man's *hetaira*, in defiance of what is right, decked out in masses of jewels and with fine rainment, going abroad in splendid state and flaunting the luxury purchased with what is ours, while they themselves are too poor to enjoy such things?
>
> (ibid., 54–5)

Here, surely, is that 'indulgence in every licence and luxury', that feminine 'greed for money'.

It is, however, in Isaios 2 [*Menekles*] that the law quoted in Demosthenes 46 is most fully appealed to. According to the account given, a certain Menekles married a woman many years his junior, the daughter of his old friend Eponymos. They produced no children and Menekles, feeling that he was depriving his wife of the happiness of a family, separated from her on friendly terms so that her brothers could arrange another marriage for her. This still left Menekles childless and without an heir, so in due course he adopted his ex-wife's brother as his son. Menekles then proceeded to live another twenty-three years. On his eventual death, however, his own brother immediately claims Menekles' estate on the grounds of being next-of-kin. Thus a contest for the inheritance arose between Menekles' brother and Menekles' adopted son – his ex-wife's brother.

The argument of Menekles' brother was, of course, that the adoption was invalid since it had been made while Menekles was under the influence of his young former wife. It is this charge which the speaker of Isaios 2 – Menekles' adopted son – must rebut, and he does so at length:

> I think, men of Athens, that if any adoption was ever made in accordance with the laws, mine was, and no one could ever dare say that Menekles adopted me in a moment of insanity or under the influence of a woman. But since my uncle, acting in my opinion ill-advisedly, is trying by every means in his power to deprive his dead brother of descendants, showing no respect for our ancestral gods or for you, I feel constrained to come to the aid of the father who adopted me, and to my own aid.
>
> (Isaios 2 [*Menekles*]: 1; Loeb trans., modified)

> That Menekles was not insane or under the influence of a woman but in his right mind you can easily understand from the following facts. In the first place, my sister, with whom most of my opponent's argument has been concerned, and under whose influence he alleges that Menekles adopted me, had remarried long before the adoption took place, so that, if it had been under her influence that he was adopting a son, he would have adopted one of her boys, for she has two.
>
> (ibid., 19)

What was at stake, of course, was an inheritance. The law relating to a woman's influence was only the technical grounds for a suit which, typically, involved the conflicting claims of collateral relatives and adopted descendants.[9] But if it was money and not Menekles' sanity with which everyone was concerned, the law relating to a woman's influence was still no dead letter. What the speaker had to establish before an Athenian male jury was still a point of fact: that Menekles was of sound mind when he had adopted him and that he had not done so under the influence of his ex-wife, for those subversive emotions which could be aroused by women were not to affect the rational and legitimate succession of Athenian *oikoi*.

V

As I hope this brief review has shown, women could be portrayed in literature – in the *Antigone* and in the *Seven Against Thebes* – as the guardians of family, kinship, and religion, a portrayal that stresses their commitment to the domestic sphere in which society situated them. Exiled from civic affairs, placed within the protection of the *oikos*, assigned the role of reproducing society as legitimate wives and mothers rather than of participating in it as citizens in their own right, they become in their 'domesticated' state the representatives and symbols of society's continuity, embodying quite literally the natural processes on which that continuity rested, and concerned with what were perceived as the equally 'natural' laws by which that continuity was sanctioned. Simultaneously, however, those same qualities, allied with nature rather than culture, with emotion rather than reason, could also be seen in literature, in philosophy, even in the provisions of Athenian law, as threatening the structure of society which, though it might have had ultimate need of them, nevertheless tried to separate itself from them. Society, culture, tolerates nature only in so far as it is necessary to it, only in so far as it can exploit it for its own survival and on its own terms, only in so far as it can domesticate it and order it. Women, perceived as part of the natural world, thus find an uneasy incorporation into society, necessary for its continuity yet psychologically unsuited to its aims.

That unease finds expression even in the most light-hearted comedy. Let me return to Aristophanes' *Thesmophoriazusai*, where Mnesilochos has infiltrated the women's festival. Disguised as a woman and speaking as a woman he relates this tale:

> I knew a woman – I won't mention names – who remained in child-birth for ten days – in order to buy a baby boy. Her husband runs to every seer buying charms to help her labour. While he's gone they bring in a baby in a crock with a honeycomb stuffed in his mouth so's he won't cry. Then, at a nod, the woman immediately cries out, 'Go away, go away, I'm about to give birth' (for the baby is kicking on the inside of the crock). The husband leaves, the wife unplugs the baby, the baby cries out. The midwife (the very one who brought the baby in) rushes out to the husband crying, 'A lion, a lion's been born you. The spitting image of your face – and everything else too. He's even got the same little prick with a twist in it like a pine-cone.' Don't we do these things? By Artemis we do.
>
> (Aristophanes, *Thesmophoriazusai*: 502–17)

Women's irrational nature and her readiness to submit to physical desire always posed the danger of seduction and adultery; of the intrusion of a *nothos*, an illegitimate, to confound the rightful succession of her husband's *oikos*. Here, however, woman's daring takes another form – a form which, it should be noted, in no way denies her commitment to hearth and home, to family life, even to pleasing her husband by the production of a son; but it does show her complete disregard for *legitimate* continuity, her complete misconception of what domesticity and family life is about – not simply the production of children (which is part of nature), but the production of children in strict accordance with the cultural requirements of civilization and the state.

I suggested earlier, however, that if by virtue of their natural characteristics women were incapable of self-governance, then the problem their incorporation into society created could be resolved by placing them under the supervision of men, who as their *kyrioi* could supply for them the rational command that they themselves were lacking. This could be construed as the solution adopted by Athenian law. Indeed, it could be construed as the solution adopted by the range of conventions and institutions – most notably marriage – that governed women's lives. But there was another option. That, quite simply, was to abandon the attempt and to leave women to wander in the wilderness of the passions which constituted their natural abode. To an extent, as we have seen, that option too was taken. There were, after all, two types of women: mothers, sisters, daughters and wives of Athenian citizens, and what might best be called the others – courtesans, prostitutes, *hetairai*, women who lay without the civic and religious structures of the state. The former, protected by law, guarded and supervised by their *kyrioi*, constrained by rigid standards of good behaviour and propriety – in short, made to conform by men to the requirements of civilized life – were in sharp contrast to the latter, performing outside the boundaries of family and *polis* no more than that to which their natural instincts fitted them: the occasional satisfaction of men's desires in an erotic relaxation from the burden of self-control. As Chaireas remarks to his love-lorn companion in Menander's comedy *The Bad-Tempered Man* (*Dyskolos*):

> In such matters, Sostratos,
> This is the way I go to work. Say one of my friends
> Is keen on a girl, asks me to help. If she's one of that sort,
> I act in a flash – I simply allow no argument;

I get drunk, burn her door down, swoop and carry her off.
Before I even ask who she is, she must be had.
The longer he waits, you see, the more he falls in love;
While if he enjoys her, he soon gets over it.
But if it's marriage he talks of, with a free-born girl,
I take a different line; enquire about her family,
Her life, the sort of girl she is; for in this case
I leave my friend a souvenir for the rest of his life
Of my own efficiency in these matters.

(Menander, *The Bad-Tempered Man*: 58–68; P. Vellacott trans.)

And if men could occasionally cross the boundary between self-control and an indulgence in eroticism, inebriation and desire, this was because they were after all men – men whose rationality would allow them to return unscathed.

Women, on the other hand, had to be placed permanently on one side or other of the divide that separated the ordered life of the *polis* from the uncontrolled world of sensuality and the passions. Either women were the chaste wives, mothers, daughters and sisters of the citizenry, or else they were creatures of pleasure. The logic of this 'double standard' is quite clear. By their very nature women belonged *as a whole* to the disordered and irrational side of life. Their integration into society was, as it were, artificial – the result of their subjugation and domestication by men. Left to themselves, women would be both ignorant of and incapable of conforming to society's demands. Those who were the chaste wives, mothers, daughters, and sisters of the citizenry might be considered not only the select, but the saved – saved by the imposition of a command which they were incapable of supplying for themselves.

But if the life of the courtesan, the prostitute, the *hetaira*, 'the other' represented in practice the second option – that of 'leaving women to themselves' – it is mythology, or rather its dramatic and poetic reworkings, that reveals the second option in its most telling form. Here an ideology of the sexes and a conception of women's nature were free to operate largely untrammelled by direct reference to social reality. Here men could imagine women not as society presented them, but as imagination (itself of course born of social conditions) could project them. In such hypothetical circumstances men could allow women free rein to exhibit to the full the 'nature' with which society endowed them. The tragic heroines of Greek drama, the divine, semi-divine, or monstrous figures of Greek mythology, may be at some considerable remove from a portrayal of the fifth- or fourth-century Athenian wife, but they are not at all distanced from the conception of female nature expressed through those laws and conventions which sought to regulate women's lives. If women were by nature wild, uncontrolled, passionate, irrational, then society had to subject them to the authority of the male; but in the indeterminant past of myth and in the collective imagination of men, women could remain free to exhibit to the full those qualities and those roles which were innately theirs. Here, then, women are presented not as they were, but as society imagined they might be. Here can be developed the threat they posed to society but which society never allowed them to realize. Here, indeed, are women fully representative of those forces which society relegated to the wilderness beyond its bounds. It is into that wilderness that it now remains to pursue them.

10

THE SAVAGE WITHOUT

I

We have seen how women were excluded from the social organization of the *polis* and how women were endowed with those characteristics considered inimical to the good conduct of a citizen, a *polites*. To a large extent social organization and psychological representations are thus homologous. Women lie outside the *polis*; the natural characteristics they possess lie outside the character of the *polites*. In myth and ritual – that is to say, in the traditional stories on which Athenian art and literature drew and the formulaic practices of worship and celebration – a third homology is apparent. For, in a variety of ways, women become located quite literally beyond the boundaries of the civilized world. They become the inhabitants of a natural, barbaric, and alien world beyond man's dominion. Conceptual space is, as it were, translated into geographical space.

In the brief remarks on Aischylos' the *Seven Against Thebes* in the preceding chapter I suggested that Eteokles identified the women of Thebes with the enemy forces drawn up against Thebes' gates, and that this identification could be interpreted in two closely connected but slightly different ways. In a quite straightforward sense Eteokles claims that the women's uncontrolled and emotional behaviour is spreading panic throughout the city and thus aiding the enemy forces outside to enslave it. At what might be considered a more symbolic level, Eteokles' speech also implies that women's excessively emotional nature is itself alien to the *polis* and represents an enslavement. Such an identification of the Theban women with those forces hostile to civilization and beyond its bounds requires a degree of interpretation of Eteokles' words, but in other contexts the identification is posited in more concrete and less ambiguous form.

As Gould notes, 'The terrifying nightmare figures of Greek mythology – the Moirai, the Erinyes, Harpies, Graiai, Sirens, Skylla and Charybdis, Medusa and the Sphinx – and bogies of folklore . . . are, again, characteristically women' (1980: 56). But they also have other characteristics in common. They all represent uncanny, uncontrollable, mysterious forces which lie beyond the civilization of men. They are part either of a particularly savage and particularly bizarre world of nature, or they are the embodiments of those strange (and perhaps divine) forces which destroy men's minds as readily as their bodies – or they are both. In this context their femininity seems apt. Women's exclusion from the *polis* and women's endowment with those characteristics inimical to ordered life are matched in myth by a representation in female form of the forces external to man's civilization and external to man's reason and threatening both. The Erinyes who, in Aischylos' *Eumenides*, sing:

> Over the beast doomed to the fire
> this is the chant, scatter of wits,
> frenzy and fear, hurting the heart,
> song of the Furies
> binding brain and blighting blood
> in its stringless melody.
>
> (Aischylos, *Eumenides*: 328–33; R. Lattimore trans.)

and who are described as:

> . . . a startling company
> of women lying all upon the chairs. Or not
> women, I think I call them rather gorgons, only
> not gorgons either, since their shape is not the same.
>
> (ibid., 46–56)

and who encounter head-on the opposition of the masculine god Apollo and the laws of Athens, are perhaps an extreme case of such a representation. The opposition between male and female, law and vengeance, reason and madness is a central theme of the play, and the Erinyes, howling on stage, present in startlingly concrete and notably feminine form those forces that civilization must overrule. But if the *Eumenides* is an extreme case, it is nevertheless appealing to a much more general and probably much less conscious association of women with what lay beyond civilization's frontiers, an association which is a recurrent theme in the Greek mythological tradition. This is the association which I shall now try to illustrate. 'The terrifying nightmare figures of Greek mythology' are but one aspect of it. More benign forms are possible and so, certainly, are more subtle ones than that presented by the Erinyes. Conscious symbolism, allegory, personification need play no part. Indeed, it is often at the simple narrative level that the Greek myths are most revealing in their recurrent association of women with a world of emotions, desires, and nature which threatens man and which man must master.

II

Although chronologically the origins of the Homeric epics are far removed from Athens and the classical city-states, Homer provides a convenient and perhaps necessary starting point for an investigation of the mythic portrayal of women. The society described by Homer – the exact status of which must remain problematic[1] – may have little connection with the Athenian *polis*, but a familiarity with Homer remained an indispensable requirement for any claim to education in the classical period. Regardless of the date of their composition, the works of Homer were 'transformed into a culturally exemplary, psychologically salient object which, once adopted by a society becomes – precisely – a myth' (Sperber 1975: 79). As such they bear directly on the *ideology* of fifth- and fourth-century Athens because of the esteem in which they were held as models of literary excellence and, more importantly, as repositories of moral wisdom.[2] Moreover, though my concern is to show a concordance between social organization and the ideologically governed perception of women, there is no need to assume a mechanical relationship between the two or to demand that they be born together of the same historical moment. Cultural traditions may possess relatively autonomous histories. The mythological (and hence literary) traditions which had been inherited from an earlier age could still supply a set of cultural associations for women which were valid within the social structure of the classical *polis*. Indeed, the proof of this is apparent in the manner in which associations, evident in

Homer, continue to reappear in the art and literature of the classical period.

In popular usage, an 'Odyssey' has come to mean a voyage of discovery and adventure. This misrepresents one important feature of the original, for inasmuch as the *Odyssey* is concerned to relate the wanderings of Odysseus, then these wanderings are part of the *Nostoi*, the 'homecomings' of the Achaians after the fall of Troy. The theme of the *Odyssey* is not a hero's voyage of exploration but his determined attempt to regain his home. As Odysseus explains in the court of the wondrous Phaiakians:

> There is Doulichion and Same, wooded Zakynthos,
> but my island lies low and away, last of all on the water
> toward the dark, with the rest far below facing east and
> sunshine,
> a rugged place, but a good nurse of men; for my part
> I cannot think of any place sweeter on earth to look at.
> For in truth Kalypso, shining among divinities, kept me
> with her in her hollow caverns, desiring me for her husband,
> and so likewise Aiaian Kirke the guileful detained me
> beside her in her halls, desiring me for her husband,
> but never could she persuade the heart within me. So it is
> that nothing is more sweet in the end than country and parents
> ever, even when far away one lives in a fertile
> place, when it is an alien country, far from his parents.
> (*Odyssey*: 9. 24–36; R. Lattimore trans., modified)

The exotic, whether represented by the Phaiakians' luxury and Kalypso's indolent isle, or by the terrors of the land of the Kyklopes and of the Laistrygones, is but an obstacle to Odysseus' return to Ithaka and his own society.

The contrast between Ithaka and the outlandish worlds through which Odysseus must pass to achieve it provides the framework for the action of the *Odyssey*. We must wait throughout the duration of four books taken up in relating affairs in Ithaka and its neighbouring states – in establishing the situation which awaits Odysseus on his return before we are introduced to the hero himself who:

> lies away on an island suffering strong pains in the palace of
> the nymph Kalypso, and she detains him by constraint.
> (ibid., 5. 13–15)

Then we are suddenly removed from the homeland of Odysseus to the enchanted island of his detainment. The tension of the epic lies in Odysseus' gradual approach towards home and towards re-entry into his society – with all the possible eventualities that the situation implies, for the Odysseus who will return is an Odysseus on whom is accreted much of the mystery and savagery of the bizarre worlds through which he has passed. More importantly, the Ithaka to which Odysseus is returning is not the Ithaka he left. It is a world in disarray, and Odysseus will not only have to regain his society but, as it were, to reinstitute it. In this context or, should I say, in this double context, women have remarkable significance.

Odysseus' image of Ithaka's domestic world is constructed around Penelope, the constant wife faithfully weaving at her loom. Equally, the greatest threat to Odysseus' return resides in the mysterious figures of such women as Kalypso and Kirke. As Kalypso enquires:

> 'Son of Laertes and seed of Zeus, resourceful Odysseus,
> are you still all so eager to go back to your own house
> and the land of your fathers? I wish you well however you do
>   it,
> but if you only knew in your heart how many hardships
> you were fated to undergo before getting back to your country
> you would stay here with me and be the lord of this household
> and be an immortal, for all your longing once more to look on
> that wife for whom you are pining all your days here.'
>
> (ibid., 5. 203–10)

To which Odysseus replies:

> 'Goddess and queen, do not be angry with me. I myself know
> that all you say is true and that circumspect Penelope
> can never match the impression you make for beauty and
>   stature.
> She is mortal after all, and you are immortal and ageless.
> But even so, what I want and all my days I pine for
> is to go back to my house and see my day of homecoming.'
>
> (ibid., 5. 215–20)

Moreover, it is around faithful Penelope weaving at her loom that the troubles which Odysseus will find on his homecoming are again centred: the suitors who would murder him and dispose of

Telemachos his son. And for all that we are continually reminded of Penelope's loyalty, her very position casts her in an ambivalent light, for in the background of the *Odyssey* there lies always the lesson of the House of Atreus and Agamemnon's fate:[3]

> Not so did the daughter of Tyndareus fashion her evil
> deeds, when she killed her wedded lord, and a song of loathing
> will be hers among men, to make evil the reputation
> of womankind, even for one whose acts are virtuous.
> (ibid., 24. 199–202)

Women are in fact fundamental to the exposition of the epic's central theme, for if the figure of Penelope is inextricable from Odysseus' goal and the domestic world of Ithaka, the other women whom Odysseus encounters are equally integral to those outlandish regions through which he must pass. And if it is Odysseus' reunion with Penelope which marks his safe return and reinstatement as Ithaka's rightful king, it is equally the activities of women that endanger this return and his successful resumption of authority. Let me briefly turn to the manner in which these women pose a threat.

It would be foolish to claim that all the dangers which Odysseus faces are women. The wrath of the god Poseidon, the anger of Helios whose cattle are slaughtered, the Kyklops Polyphemos, the Laistrygones, and the monsters Skylla and Charybdis (though the last are, in a sense, female) all take their toll of Odysseus' men and diminish his chances of return. But it should be noted that the dangers which Odysseus faces are of two sorts: some are purely physical, others more insidious. The Land of the Lotus-Eaters comes immediately to mind, for the lotus destroys men's resolve, not their bodies, and it is in this category of dangers that the *Odyssey*'s women can for the most part be classed.

Kalypso is no frightful figure; she becomes Odysseus' saviour, friend and lover. But it is also she who proves the greatest obstacle to his return, or at least it is she who succeeds in detaining Odysseus from Ithaka for the longest time. As Athena complains to Zeus:

> she detains the grieving, unhappy
> man, and ever with soft and flattering words she works to
> charm him to forget Ithaka; and yet Odysseus

straining to get sight of the very smoke uprising
from his own country, longs to die. (ibid., 1. 55–9)

Some might envy Odysseus his fate; but as for Odysseus himself:

his eyes were never
wiped dry of tears, and the sweet lifetime was draining out
of him,
as he wept for a way home, since the nymph was no longer
pleasing
to him.
(ibid. 5., 151–4)

Odysseus needs his resolve, for what Kalypso offers is not only her own ageless beauty, but immortality for himself in an endless life of sensual pleasure. Yet Odysseus chooses mortality and prefers Penelope, for Penelope represents something more important than sensual pleasure; she is part of that close network of family, kin, and friends that forms the social environment without which Odysseus has no identity as a man. The pleasure Kalypso offers can be enjoyed only at the expense of being cut off from society in a luxurious world which nevertheless lies outside the civilization of men.

Kirke is not dissimilar. To be sure, the first encounter is not promising: half Odysseus' men are turned into swine and Odysseus himself avoids the same fate only because he is armed with Hermes' magic and advice. But, true to Hermes' words, once Odysseus has drunk Kirke's potion without harm and rushed at her with drawn sword she falls at his knees and pleads:

Come then, put away your sword in its sheath, and let us
two go up into my bed so that, lying together
in the bed of love, we may have faith and trust in each other.
(ibid., 10. 333–5)

Odysseus extracts an oath of good behaviour from Kirke, avails himself of her invitation, and his men are restored to their human shape while Kirke thenceforth acts as an admirable hostess. Kirke is transformed from an enemy to a faithful lover; but again it should be noted that her sexuality and the bounty she provides themselves continue to threaten Odysseus' purpose. It is his men who must remind him of Ithaka, for:

> There for all our days until a year was completed
> we sat there feasting on unlimited meat and sweet wine.
> But when it was the end of a year, . . .
>
> . . .
>
> then my eager companions called me aside and said to me:
> 'What ails you now? It is time to think about our own country'
> (ibid., 10. 467–72)

In the land of the Phaiakians, Odysseus actually claims that:

> Aiaian Kirke the guileful detained
> me beside her in her halls, desiring me for her husband.
> (ibid., 9. 31–2)

Nor should what might appear an unimportant detail pass unnoticed: before Kirke changed Odysseus' men into swine by striking them with her wand she gave them:

> malignant drugs, to make them forgetful of their own country.
> (ibid., 10. 236)

The precaution may seem unnecessary, but it does suggest Kirke's classification with those dangers which, like the lotus, do not simply destroy men but distract and enslave their minds.

Even the innocent Nausikaa, daughter of Alkinoos and princess of the Phaiakians, endangers Odysseus' return. Odysseus' encounter with Nausikaa is a delightful episode, almost a romantic interlude in the tale of his struggles – and Odysseus' behaviour is impeccable. The scene is presented as a model of social propriety. Surprising Nausikaa and her maidens by the shore:

> great Odysseus came from under his thicket,
> and from the dense foliage with his heavy hands he broke off
> a leafy branch to cover his body and hide the male parts.
> (ibid., 6. 127–9)

Having supplicated Nausikaa, by whom he is kindly received, Odysseus is given oil to anoint himself and directed to a stream where he can wash. Once more he is overcome by modesty:

> 'Stand as you are, girls, a little way from me . . .
> I will not bathe in front of you, for I feel embarrassed
> in the presence of lovely-haired girls to appear all naked.'
> (ibid., 6. 218–22)

Nausikaa, for her part, is equally careful to respect social conventions. She will direct Odysseus to the city and the palace of her father, but she and Odysseus are to arrive there separately for:

> 'it is their graceless speech I shrink from, for fear one may
> mock us
> hereafter, since there are insolent men in our community,
> and see how one of the worse sort might say when he met us,
> "Who is this large handsome stranger whom Nausikaa
> has with her, and where did she find him? Surely he is
> to be her husband . . ."
>
> So they will speak, and that would be a scandal against me,
> and I myself would disapprove of a girl who acted
> so, that is, without the good will of her dear father
> and mother making friends with a man, before being formally
> married.'
>
> (ibid., 6. 273–89)

Yet if the decorum and modesty of this scene is in marked contrast to the blatant sensuality of Odysseus' encounter with Kalypso and Kirke, still the whole situation is infused with eroticism. Nausikaa's ripeness for marriage is constantly stressed; her nubile beauty and that of her attendant maidens is very much to the forefront of the scene. Secreted in the bushes, Odysseus watches as:

> they all threw off their veils for a game of ball, and among them
> it was Nausikaa of the white arms who led the dancing;
> and as Artemis. . .
> she is easily marked among them, though all are lovely, so this
> one shone among her handmaidens, a virgin
> unwedded.
>
> (ibid., 6. 100–9)

Nor does Odysseus' address to Nausikaa hide his appreciation of the spectacle:

> I must find you in the nearest likeness to Artemis
> the daughter of great Zeus, for beauty, figure and stature. . . .
>
> Wonder takes me as I look on you.
> Yet in Delos once I saw such a thing, by Apollo's altar.
> I saw the stalk of a young palm shooting up . . .
>
> (ibid., 6. 151–63)

Moreover, with a little help from Athena, Odysseus, too, can appear the suitor:

> taller,
> for the eye to behold, and thicker, and on his head she arranged
> the curling locks that hung down like hyacynthine petals . . .
> and he went a little aside and sat by himself on the seashore
> radiant in grace and good looks; and the girl admired him.
>
> (ibid., 6. 229–37)

It would not, I think, be out of place to suggest the existence of a variant in which Odysseus actually married Nausikaa.[4] As the story stands, however, he does not, but instead takes a tender leave and frees himself from the temptations – temptations which are again both hedonistic and erotic – which would hold him back from society, friends, kin, and wedded wife. For the land of the Phaiakians may be a pleasant and ordered society, but it is still a magical world beyond the ken of ordinary mortals.

Finally there is the allure of those strange female monsters, the Sirens. As Kirke warns Odysseus:

> You will come first of all to the Sirens, who are enchanters
> of all mankind and whoever comes their way; and that man
> who unsuspecting approaches them, and listens to the Sirens
> singing, has no prospect of coming home and delighting
> his wife and little children as they stand about him in greeting,
> but the Sirens by the melody of their singing enchant him.
>
> (ibid., 12. 34–44)

It is only by adopting Kirke's ruse of plugging the ears of his men with wax and tying himself to the mast of his ship that Odysseus escapes their call. As Sokrates was to comment in the classical period:

> I am convinced that the reason why men fled Skylla was that she laid hands on them; but the Sirens laid hands on no man; from far away they sang to all and therefore, we are told, all submitted and on hearing were enchanted.
>
> (Xenophon, *Memorabilia*: 2.6.31; Loeb trans., modified)

It was the very beauty of the Sirens' song which lured men to their destruction. Like Kalypso, Kirke, and even Nausikaa, it is the attractions the Sirens offer, not their horror, which would prevent Odysseus' return to civilization.

Kalypso, Kirke, Nausikaa, and the Sirens are, of course, by no means identical figures. Some are malignant, some benevolent. The Sirens are indisputably evil; Nausikaa the ideal of the modest maiden. Nevertheless, all these females threaten to prevent Odysseus' return in essentially the same way, for whether they are sexually avid or innocently attractive they represent a sensuality that is opposed to domesticity and to society. But they contrast with the domestic ideal not only by being located in a strange world beyond the bounds of civilization where erotic attraction might tempt Odysseus to remain. In one way or another they are all themselves creatures of the wild, inhabitants of a world of nature with which they have strong affinity and over which they exercise command.

Kalypso's island is an idyllic reserve, her own domain. She is, after all, a nymph – one of those semi-divine, youthful, and immortal females who populate the world of Greek mythology as the guardians of its fountains, rivers, and woods. The initial description of Kalypso stresses her natural surroundings. As Hermes goes to seek her out:

> after he had made his way to the far-lying island,
> he stepped then out of the dark-blue sea, and walked on over
> the dry land, till he came to the great cave, where the lovely-
>   haired
> nymph was at home, and he found that she was inside.
>   There was
> a great fire blazing on the hearth, and the smell of cedar
> split in billets, and sweetwood burning, spread all over
> the island. She was singing inside the cave with a sweet voice
> as she went up and down the loom and wove with a golden
>   shuttle.
> There was a growth of grove around the cavern, flourishing,
> alder was there, and the black poplar, and fragrant cypress,
> and there were birds with spreading wings who made their
>   nests in it,
> little owls, and hawks, and birds of the sea with long beaks
> who are like ravens, but their work is on the sea water;
> and right above the hollow cavern extended a flourishing
> growth of vine that ripened with grape clusters. Next to it

> there were four fountains, and each of them ran shining water,
> each next to each, but turned to run in sundry directions;
> and round about there were meadows growing soft with
> parsley
> and violets, and even a god who came into that place
> would have admired what he saw, the heart delighted
> within him.
>
> (*Odyssey*: 5. 55–74)

Kalypso resides at the centre of a profusion of nature. So too does Kirke, whose halls lie through the undergrowth and forest on her island. But Kirke not only lives in the midst of nature, she also commands it:

> In the forest glen they came upon the house of Kirke . . .
> all about it there were lions, and wolves of the mountains,
> whom the goddess had given evil drugs and enchanted,
> and these made no attack on the men, but came up thronging
> about them, waving their long tales and fawning.
>
> (ibid., 10. 210–15)

Indeed, Homer's depiction of Kirke makes her strongly resemble the *Potnia Theron*, the 'Mistress of Animals', known from the Mycenaean period and a favourite subject of archaic art (Nilsson 1964: 28). This concept of a female divinity ruling over the inhabitants of the natural world continues into the classical period with the figure of Artemis on whom the title *Potnia Theron* is also bestowed and whom the chorus of Aischylos' *Agamemnon* address:

> Lovely you are and kind
> to the tender young of ravening lions.
> For sucklings of all the savage
> beasts that lurk in the lonely places you
> have sympathy.
>
> (Aischylos, *Agamemnon*: 140–4; R. Laltimore trans)

But Kirke's affinity with animals goes further. Odysseus' men are turned by her magic into swine. They are transmogrified into the beasts over which this *Potnia Theron* rules. To enter into Kirke's domain is to become victim of her power, one of the animals over which she rules.

Nor is it insignificant that Odysseus encounters even Nausikaa and her maidens in a rural setting. He spies on the young girls gambolling together without their veils, and their similarity to Artemis and her nymphs is immediately developed:

> and as Artemis, who showers arrows, moves on the mountains
> either along Taygetos or on high-towering
> Erymanthos, delighting in boars and deer in their running,
> and along with her nymphs, daughters of Zeus of the aegis,
> range in the wilds and play, and the heart of Leto is gladdened,
> for the head and the brows of Artemis are above all the others,
> and she is easily marked among them, though all are lovely,
> so this one shone among her handmaidens, a virgin unwedded
> (*Odyssey*: 6. 102–9)

In fact Nausikaa is not only Artemis the virgin goddess of the woods but also 'the stalk of a young palm-tree shooting up', and interestingly as Odysseus himself approaches Nausikaa and her maidens he too becomes a denizen of the wild – a lion stumbling into an innocent flock (*Odyssey*: 6. 130–8).

Finally, with the Sirens singing in their meadow, we encounter one of those common mythological hybrids, part-human, part-beast in form, and though Homer offers no physical description of the Sirens, they appear later on carved funeral monuments as birds of prey – similar, indeed, to those other winged female predators the Harpies or Snatchers (Smith 1892).

Kalypso, Kirke, Nausikaa, and in somewhat different form, the Sirens are all, then, seductresses – at least inasmuch as all offer the promise of a form of sensual pleasure which would detain Odysseus from his return to domesticity and ordered life. And all, since they belong to the outlandish and fantastical world of his travels, lie by definition beyond the frontiers of civilization. Finally, all are presented by the poet, whether through specific attributes or through the imagery by which they are described and the context in which they appear, as part not only of an exotic world but of a world of nature.

By contrast, Penelope, Odysseus' wedded wife, the mother of his son, is central to the vision of a domesticity which Odysseus retains, and with whom his reunion will herald civilization regained. Yet strangely the other women whom Odysseus encounters on his travels not only contrast with Penelope; they also evoke her presence. Like Penelope, Kalypso is weaving at her loom and the sweet smoke rises from her hearth. Kirke, too, appears in a similar setting. As Gould comments, 'Nothing, we know, is as it should be, and yet everything is familiar' (1980: 53). We are presented with women who seem radically different, and yet we are continually reminded of basic similarities. It is as if all are possible variations of some underlying unity. Their similarities and their differences seem to map out a typology of combined and opposed characteristics into which many of the mythical representations of women will fit. Between Kalypso and Kirke on the one hand and Penelope on the other an opposition is created between woman as a creature of the wild and woman as the model of domesticity. They represent the extremes of a scale both geographically and with respect to their social and sexual roles. Nausikaa, however, is an intermediary figure, and one which suggests a continuity which extends from Kalypso and Kirke to Penelope, for Nausikaa is situated precariously between the wild and the domestic, between nature and culture. As an umarried virgin she is, like Kalypso and Kirke, a creature of the wild. She is Artemis or one of her nymphs dancing by the stream, or she is the stalk of a young palm tree shooting up. But her encounter with Odysseus also stresses her adherence to the conventions of society. With marriage before her she is soon to be led home by some man to become, like Penelope, a faithful wife and the mother of children. If Kalypso and Kirke are placed at some distance from Penelope, it is a distance which Nausikaa will travel in the course of her maturation. The continuity is there and the wedded wife, the Penelope, is not a different species from Kalypso or Kirke. She is not domesticated woman permanently opposed to the women of the wilds; rather, she is woman in her domesticated state, the end product into which Nausikaa will turn upon her marriage and incorporation into the household of a man.

III

As I have already stressed, the society described in the Homeric epics cannot be equated with that of fifth- and fourth-century Athens. The manner in which women and female figures are presented does, however, form part of a continuing cultural tradition reiterated throughout the Greek mythological corpus and apparent in the Athenian and classical renditions. As I have also stressed, there is no question of the *Odyssey* (or of any other myth to be reviewed) having to be read as an allegory in which certain ideas about the social position of women or their psychological characteristics are consciously translated into a series of concrete situations and representations. The *Odyssey* can be taken quite simply as the tale of Odysseus' wanderings in his attempt to return to Ithaka. But inherent in such a tale there is nevertheless a certain manner of representing women, a certain series of situations and roles in which they habitually appear, making them the natural inhabitants of the wild, dangerous and mysterious regions which lie beyond the borders of civilization. The *Odyssey* is not, I think, thereby concerned to comment on the nature of women, nor is it allegorically representing the dangers and attractions of the wild in the form of women. But as part of the ordering process inherent in any human representation of the world it nevertheless contains within it a series of classifications and associations which cast them in that light.

On the basis of Homer alone the interpretation just given of the figure of Nausikaa might appear strained, but the language of metaphor as it related later in the classical period to women, sex, and marriage both reinforces the interpretation and shows the continuity. As Gould notes, 'The formula required by custom if not by law, for giving in marriage is several times attested in Menander:[5] "I give (*enguo, didomi echein*) you this woman (my daughter) *gnesion paidon ep'arotoi*", "for the ploughing of legitimate children" ' (1980: 53). As Gould argues, 'This traditional formula is part of a network of imagery and metaphor which associates women and their role in sex and marriage with animals, especially the taming, yoking and breaking in of animals, and with agriculture.' Marriage is a yoke placed on virgins who are wild or untamed – who are unbroken horses or bitches; or else women are land to be ploughed or fields to be sown.[6]

That an agricultural society should use agricultural imagery for sexual activities (or for anything else) is not surprising; but such a 'network of imagery and metaphor' can still be taken as betraying a particular attitude towards women. However 'natural' a system of metaphors may seem within a particular historical context, metaphors are still part of the *cultural* process whereby a society apprehends itself and the world. Women are fields to be cultivated, ploughed, sown; animals to be broken in, mounted, domesticated. The metaphors of an agricultural society are agricultural, but much about gender relations is still revealed, for it is men who are consistently identified with the activities of culture itself, while women remain part of nature, part of the wild, which men must 'cultivate' (cf. Gould 1980: 57).

In myth and in Athenian ritual this imagery is extensively developed. At Brauron, the sanctuary of Artemis Brauronia on the east coast of Attica some 38 km from Athens, Athenian girls between the ages of five and ten participated in a festival in which, dressed in yellow robes, they 'did the bear' by imitating small bears walking on their hind legs.[7] As with so many of the activities connected with Athenian festivals, the precise details of the ritual remain obscure, but an admittedly late source[8] explains that this mimicry was in preparation for their future marriage; that by imitating bears they would rid themselves of all their savagery, shedding their wild state along with their yellow robes which they removed in the course of the ritual. Artemis' little wild animals prepare to become the wives of the citizens of Athens (cf. Tyrrell 1984: 73–4).

For its part, myth presents us with a series of women – of whom the goddess Artemis herself is the perfect representation, but among whom must be numbered Persephone, daughter of Demeter, worshipped under the name simply of *Kore* 'the Maiden', Atalanta, even Nausikaa, and those countless nymphs of the forests, springs, and streams who in their virginal state are all creatures of the wild. Society, on the other hand, presents us with women who in their marital state are domesticated and incorporated into society as the wives of men. Ritual effects the transition from the (imaginary) wildness of myth to the desired (and real) domestication of society, and it should be remembered that the word *nymphe* denoted not only the semi-divine and immortal young female who presided over the mountains, forests, streams, or springs of the countryside, but also any marriageable young maiden or bride. In the metaphors of common speech women are horses to be broken in, fields to be ploughed: in the representations of mythology, then, as Vernant says, 'Shunning contact with males, living far from men and the life of the city, the *kore* [maiden], like Artemis, the virgin huntress, mistress over wild animals and uncultivated land, shares in the life of the wild' (1982: 139). A woman's role as mother and wife is made consequent on her capture and domestication by men.

The ritual expression of this domestication is, of course, apparent in the ceremony of marriage itself – a ceremony which transferred the young bride, the *kore*, the *nymphe*, from the province of Artemis the virgin huntress to that of another female divinity, Demeter. And the agricultural metaphors – or rather, the classifications of status which reflect the opposition between nature and culture, the wild and the domesticated – are continued; for, in contrast to Artemis, Demeter is not only the patroness of married women, she is also the goddess of agriculture, of crops and cultivated land. As Detienne argues, 'under the sign of Demeter, marriage turns out to be inseparable from the life of cultivated plants' (1977: 116), for beyond the traditional marriage formula in which a woman is identified with a field to be sown by her husband, the ritual of marriage recalls in three further ways the introduction of cultivation. First, on their marriage day Athenian girls were required to carry a pan for roasting barley; second, a pestle for grinding corn or barley was hung outside the marriage bed-chamber while the sieve which went with it was carried by a young child in the wedding procession; finally, on the wedding day a child wearing a crown of thorny plants entwined with acorns offered bread from a winnowing basket to all the guests saying, 'I have fled from evil and found what is best', which was interpreted within Greek tradition as a reminder of the transition from the 'thorny life' (*bios akanthodes*) to the 'cultivated life' (*bios hemeros*), from the 'wild life' (*bios agrios*) to the 'life of milled corn' (*bios alelesmenos*) (Detienne 1977: 117). As Detienne concludes:

> Set in such a context, the various culinary instruments paraded on the wedding day – the pan for roasting barley, the sieve and the pestle – can be seen as mediators between the two extremes reconciled in the figure of the child who is both crowned with thorny plants and loaded with bread. At one extreme we have the wild fruits and plants which represent the food of an age that preceded the cultivation of cereals; at the other, the loaves of bread ready to be eaten as a pledge for the newly married couple's impending participation in the life of milled corn.
>
> (Detienne 1977: 117)

The transition of the girl from virgin to matron is made ritually parallel in the wedding ceremony with the transition of mankind from a pre-agricultural to an agricultural age. Both transitions involve the domestication of nature and a shift from the province of Artemis to that of Demeter.[9]

But if in society this domestication of the female could regularly and successfully be achieved through the institution of marriage and the rituals which accompanied its inception, in the imaginative constructs of mythology the task appears not so easy. The stories of Atalanta illustrate the attendant dangers.

According to the account of the Hellenistic mythographer Apollodoros,[10] Atalanta's father, Iasos, had wished for a son. Disappointed by the birth of a daughter, he exposed Atalanta on a hill where she was suckled by a bear until later found by hunters who reared her. She carried arms, became a hunter herself, and remained a virgin; she became, in fact, a sort of mortal Artemis. Eventually she was recognized by her father who immediately set about arranging a marriage for his reluctant daughter. But Atalanta imposed a condition: any suitor for her hand had either to beat her in a foot race or be killed. Many young men lost their lives. Finally, she was overtaken and claimed by a certain Melanion, but only because with Aphrodite's assistance and on her advice Melanion employed the ruse of casting before him golden apples which Atalanta stooped to pick up. The marriage took place but was short-lived, for one day Melanion persuaded Atalanta to sleep with him in a sanctuary of Zeus. Outraged by the defilement, Zeus changed them both into lions. And here it is worth noting an interesting detail. According to later mythographers this curious punishment was chosen because lions do not mate with each other, but only with leopards.[11]

There are other treatments of the myth, but its essential features remain the same:[12] Atalanta, like Artemis, is a virgin, a huntress, a denizen of the wild. Her virginity and her wildness are inseparable, and she stands to lose both on marriage. Like a wild animal she must be pursued and caught by the man who would domesticate her; and that man runs the risk of all ambitious hunters – to be killed by his quarry. Atalanta is caught; but in the long term her domestication proves unsuccessful. Both she and her husband are returned to nature, transmogrified into lions. And, if the later mythographers are taken into account, then Atalanta's return to a state of nature also marks a return to sexual abstinence for, as a lion, she can no longer have sexual intercourse with her mate.

In fact the capture or abduction by some man (or god) of the maiden or nymph from her natural world is one of the recurrent motifs of Greek mythology – nor is there any necessity to speculate about a folk memory of some hypothetical primitive past in which 'marriage by capture' was the norm in order to explain this. The motif relates not to some lost historical institution, but to the actual classifications of culture: woman must be captured and domesticated to become part of society. Prominent among such stories is that of 'the Maiden' herself, Kore or Persephone, who, in the Homeric *Hymn to Demeter*, is seized by Hades while she is gathering flowers in a meadow and carried off to the underworld to become its queen; or the capture by Peleus of the sea nymph Thetis who, to avoid being taken, changed variously into fire, water, a tree, a bird, a tiger, a lion, a serpent, and a cuttlefish. Again, the domestication of neither was entirely successful. Peleus obtained an immortal bride who was to become the mother of the hero Achilleus, but Peleus himself, a mortal, was doomed to age and Thetis finally left him; Hades was permitted to keep his bride for only one third of the year, her partial release from marriage and subjugation being secured by her mother, Demeter.[13] It is, however, Artemis herself who constitutes the paradigm of virginity in its immutable state, and it is Artemis who most clearly reveals the connection between virginity and the life of the wild.

Martin Nilsson, the great historian of Greek religion, wrote:

> Artemis was the most popular goddess of Greece, but the Artemis of popular belief was quite a different person from the proud virgin of mythology, Apollo's sister. Artemis is the goddess of wild Nature, she haunts the mountains and woods, the groves and luscious meadows. There the 'rushing Artemis' (*keladeine Artemis*) hunts and dances with her attendant nymphs. She protects and fosters the young of animals and growing human children. Different animals are her companions and theriomorphic representations are appropriate to her. In her cult occur orgiastic dances and the sacred bough.
> (1964: 28)

Presumably Nilsson found the wilder, ecstatic aspects of Artemis' role in cult and popular belief to be at odds with the austere figure of chastity which he derived from literature. However, the portrayal of Artemis as an ecstatic goddess of nature is not entirely absent even from sophisticated literary works, for at least in Euripides' *Hippolytos*, Artemis, far from representing a pure and passive chastity in her opposition to eroticism instead combines sexual abstinence with a considerable savagery. Indeed the 'intriguing polarity', as Kirk (1974: 62) calls it, between Artemis and Aphrodite, goddess of 'love', must be carefully examined.

The conflict between Artemis and Aphrodite, between chastity and eroticism, is clear in Greek mythology. The two goddesses are traditional enemies, the qualities they represent self-evidently opposed. Recall that Atalanta's undoing was caused by Aphrodite. But in many respects their opposition is not so marked – or rather, both Aphrodite and Artemis pose a common threat to the ideal of domesticated woman. Let me turn briefly to the *Hippolytos*, for there both Aphrodite and Artemis stand as savage forces between which the luckless mortals, Hippolytos and Phaidra, are crushed.

Aphrodite's appearance opens the play. In a speech which makes clear its whole development she states with chilling arrogance the cause of her anger and announces Hippolytos' imminent death:

> Hippolytos, son of Theseus by the Amazon
> pupil of holy Pittheus,
> alone among the folk of this land of Troezen has blasphemed me
> counting me vilest of the Gods in Heaven.
> He will none of the bed of love nor marriage,
> but honors Artemis, Zeus' daughter,
> counting her greatest of the Gods in Heaven
> he is with her continually, this Maiden Goddess, in the greenwood.
> He hunts with hounds and clears the land of wild things,
> linked in companionship too high for a mortal.
> I do not grudge him such privileges: why should I?
> But for his sins against me
> I shall punish Hippolytos this day.
> (Euripides, *Hippolytos*: 11–22; D. Grene trans., modified)

In turn Artemis' appearance marks the virtual close of the play, and in a speech in which she mercilessly explains to Thesus how he has wrongfuly cursed and killed his own son, she nevertheless condescends to give comfort:

> You have sinned indeed, but yet you may win pardon.
> For it was Kypris [Aphrodite] managed the thing this way
> to gratify her anger against Hippolytos.
> This is the settled custom of the Gods:

No one may fly in the face of another's wish:
we remain aloof and neutral. Else, I assure you
had I not feared Zeus, I never would have endured
such shame as this – my best friend among men
killed, and I could do nothing.

(ibid., 1325–33)

To the dying Hippolytos, however, she explains that his death will not go unavenged:

Hush, that is enough! You shall not be unavenged,
Kypris shall find the angry shafts she hurled
against you for your piety and innocence
shall cost her dear.
I'll wait until she loves a mortal next time,
and with this hand – with these unerring arrows
I'll punish him.

(ibid., 1416–20)

We are left with the knowledge that Artemis, and Aphrodite's quarrel will be pursued at the expense of further victims, that this battle of the gods, this contest between the powers of absolute chastity and absolute eroticism will continue to consume those unfortunate enough to be caught in their opposing claims. Within the play itself, however, Aphrodite works her revenge through the person of Phaidra. Despite her genuine attempts to control the passion by which she has been afflicted, Phaidra is driven to the point of madness by love for Hippolytos, her husband's son. Once her passion is revealed, then, driven by guilt and shame, she suicides – but not without ensuring Hippolytos' death, for in a tablet clutched in her hand and addressed to Theseus her husband she falsely accuses Hippolytos of rape.

Phaidra's susceptibility to *eros* and the plot she contrives conform to the stereotype of feminine character and action outlined in the preceding chapter. Phaidra's passion again proves the danger of introducing into society women and the emotions to which they are subject: father curses son, who is horribly killed; Theseus is left with his house destroyed and without an heir. But if this is the work of Aphrodite and her mechanic, it is also made clear in the play that the Artemis whom Hippolytos worships and the way of life which he adopts in imitation of her are equally opposed to the civilized ideals of self-mastery and restraint. Hippolytos is, of course, male, and his violent hatred of all women expressed at length in one of the play's central speeches is the best known anti-feminine tirade of Greek literature (Pomeroy 1975: 106):

> Women! This coin which men find counterfeit!
> Why, why, Lord Zeus, did you put them in the world,
> in the light of the sun? If you were so determined
> to breed the race of man, the source of it
> should not have been women. Men might have dedicated
> in your own temples images of gold,
> silver, or weight of bronze, and thus have bought
> the seed of progeny . . . to each been given
> his worth in sons according to the assessment
> of his gift's value. So we might have lived
> in houses free of the taint of women's presence.
>
> (ibid., 616–24)

Hippolytos would do without women altogether. Male offspring for the continuity of male society could be achieved by more rational means. Indeed, the question 'why women at all?' seems to have been one that long exercised the Greek imagination, for it was dwelt on (and answered) by Hesiod. Hippolytos appears to accept Hesiod's conclusion, that women are a curse, an evil, a deception brought upon mankind in punishment; that with the end of the golden age, just as man must labour with his body to eat so he must endure women in order to reproduce.[14] But given that men must endure women, that the continuity of society is dependent on the domestication of women, Hippolytos' total rejection of women, sex, and marriage takes him once more beyond the bounds of civilization and into the territory of what, it must be stressed, is another female deity. Hippolytos rejects the passion of Aphrodite only to become the passionate follower of an equally unrestrained goddess, Artemis. He has indeed 'kept company too high for a mortal', too single-minded for civilization and its survival, for both Aphrodite and Artemis represent female extremes outside its requirements, one by an excessive eroticism that cannot be contained within marriage, the other by an excessive chastity that will not allow marriage.[15]

At the level of myth, in the figure of Aphrodite and Artemis, these extremes remain immutable, forever opposed. For society, that could not be, and it was between the extremes of virginal wildness and passionate eroticism that society had conceptually to situate those women whom it was willing to take into its fold. Their patroness was Demeter, sexually fecund, but non-erotic – and here we return to that network of imagery which associates the role of women in sex and marriage with agriculture. Artemis and Aphrodite do not represent a simple opposition; rather, with Demeter, they constitute a triad, and both Artemis and Aphrodite, one by excessive chastity, the other by excessive eroticism, are opposed to woman as society required her, domesticated and productive, fertile yet chaste. As Vernant (1982: 139) argues, on entering into marriage, the required and desired state for every Athenian woman, a young girl enters the domain that belongs to the deity of cereals. Here she must rid herself of all the wildness inherent in the female sex, a wildness which can take two opposed forms, for it might make the woman veer towards Artemis, falling short of marriage and refusing any sexual union, or, on the other hand, it might propel her in the other direction, beyond marriage, towards Aphrodite and into unbridled erotic excess. 'The position of the *gyne enguete*, the legitimate wife, is in between that of the *kore*, the young girl defined by her virginal status and that of the *hetaira*, the courtesan entirely devoted to love' (Vernant 1982: 139).

The ritual opposition in Athens between the domesticated sexuality of Demeter and the 'erotic excess' of Aphrodite has been extensively studied by Detienne in *The Gardens of Adonis* (1977). The *Thesmophoria*, the great women's festival of Athens sacred to Demeter, was an official part of Athenian religious life. The *Adonia*, in honour of Aphrodite's lover, Adonis, was 'an exotic festival tolerated by the Athenian city on the periphery of the official cults and public ceremonies, a private affair' (1977: 65). The *Thesmophoria* was reserved exclusively for married women, the legitimate wives of citizens. The *Adonia* was celebrated principally by prostitutes, courtesans, and their lovers. The *Thesmophoria* strictly excluded all men, and for the duration of the festival Athenian matrons also abstained from sexual intercourse with their husbands. The *Adonia* was marked by sexual licence and revelry.

The most intriguing feature of the festival of Adonis, however, was the custom of planting cereals in small pots which were placed on the roof-tops of houses. Exposed to the intense heat of summer, these plants immediately sprouted only to wither and die within a matter of days. The plants of Aphrodite and her lover, Adonis, are ultimately sterile, and between the *Thesmophoria* and the *Adonia* there is thus not only a contrast between legitimate wives and prostitutes, between marriage and seduction; in terms of what Detienne calls the 'vegetable code' there is also a contrast between the fertility and growth of crops presided over by Demeter, and the precocious sprouting and withering away of the plants of Aphrodite and Adonis. In ritual the over-evaluation of sexuality and the erotic is coupled not with fecundity, but with barrenness. By contrast, those women who are fertile and who will produce legitimate children celebrate a festival which is marked by sexual abstinence and a rejection of the erotic. In this context two further details need to be noted. During the festival of the *Thesmophoria* the women used the branches of a certain type of willow, the *agnos*, to make couches on which to rest and took on the ritual name of *Melissai*, the 'bees' of Demeter (Detienne 1977: 79–80). Within Greek tradition this type of willow was held to possess anti-aphrodisiac qualities, while bees, the very model of female domestic virtue (and precisely the one which Ischomakhos proposes for his young wife in Xenophon's *Oikonomikos*), were notable not only for their industry, but also for the asexual manner of their reproduction (cf. Pomeroy 1975: 49).

It was Demeter, then, who stood opposed to both Artemis and Aphrodite. It was woman separated from the wildness of both virginity and of eroticism who represented the social ideal. Through the figure of Demeter, woman's marital fertility and her marital chastity were associated with the cultivation of plants and crops, with culture as opposed to nature (cf. Vernant 1982: 139). What both Artemis and Aphrodite stand for lies outside the requirements of civilization, and in many myths the theme of woman's affinity with nature and the wild is not opposed to, but combined with erotic desire. To revert to Homer: Kirke and Kalpso, like Nausikaa, are presented as part of a world of nature, yet in contrast to Nausikaa (the virgin who will take the correct path towards domestic life) they are sexually avid. And it is worth noting the existence of a version of the Phaidra and Hippolytos story in which Phaidra claims to have been converted to the cult of Artemis and suggests to Hippolytos that he in turn should pay homage to Aphrodite in her company.[16] Indeed, in many myths the coupling of nature and of erotic desire is taken to a literal extreme. Pasiphai, the wife of King Minos of Crete, is overcome by desire for a white bull and, with the aid of Daidalos and the construction of a hollow wooden cow, consummates her desire – the result being, of course, the dreadful minotaur, half-bull, half-man. The nymph, Leda, is seduced (or raped) by Zeus in the form of a swan. Europa, playing in the meadows by the sea, is abducted by Zeus in the form of a white bull.[17] The copulation of women with animals is not an uncommon theme.

Unlike Pasiphai, however, it would seem that neither Leda nor Europa were the willing victims of their animal lovers, and such differences should be noted, for I am not suggesting that Greek myths were so systematically interrelated that they reveal any single and unalterable pattern in their presentation of women. Nor, on the other hand, do I wish to suggest that the differences themselves form part of some complex series of transformation in which the 'inversions' as much as the reiterations of certain relationships should be seen as the workings of some underlying and abstract Lévi-Straussian logic. Nevertheless, certain themes that recur in a variety of contexts, in a variety of combinations, and with a variety of effects do suggest in a relatively unsystematic and fluid fashion an habitual association of women with various aspects of what was seen to lie outside the society of men. Women are mythically located in a world of nature, with which they have strong affinity either as its recalcitrantly wild or its vulnerably innocent inhabitants, in which like Artemis or Atalanta they revel or from which, like Europa, Leda, or even Persephone, they are abducted; or else they are made the victims or the bearers of those uncontrollable desires and passions from which, equally, society had to distance itself (cf. Gould 1980: 52). But whether passionate or chaste, wild or innocent, it is men who must domesticate women to make them mothers and wives. And, if we turn to the figure of the Amazon, it can be seen that this set of associations is imaginatively reproduced and elaborated not only in myth, literature and ritual but, in what could pass for history, entered into the political and moral homilies of Athenian oratory.

IV

The homeland of the Amazons is variously given: in Lycia and in Phrygia; to the north of Greece in Thrace and as far east as the Caspian; along the coast of Asia Minor; and, most commonly, on the southern shore of the Black Sea on the Thermodon river (Tyrrell 1984: 56; Engle 1942: 513–14).[18] Portrayed in mythology as an independent society of warrior women, either they had husbands who were kept in strict subordination, or they mated promiscuously with men of neighbouring tribes,[19] the female children being kept, the male children being variously returned to their fathers' tribe, or mutilated to provide harmless cripples.[20] In many accounts the right breast was seared or removed so that it would not interfere with the use of weapons, a practice which furnished the usual Greek derivation for 'Amazon' i.e. *amazos*, without a breast.[21] They are never portrayed in this fashion in pictorial representations, however (Engle 1942: 517), and according to Philostratos, they fed their female infants on horse's milk to prevent the enlargement of their breasts. 'Amazon' thus meant 'unsuckled'.[22] They were themselves horsewomen, and are frequently depicted as such.[23] Above all, they were ferocious warriors and, as Apollonios put it, they were no gentle folk, cared not for justice, and were intent on violence and the works of Ares.[24] Finally, as their most frequent epithets emphasize – *antianeira* (striving-against-men), *deianeira* (man-murdering), *androphonos* (slayer-of-men), *androktonos* (man-slaying) etc. – they were fundamentally hostile to men (Engle 1942: 517–18).

Whether stories of the Amazons reflected any historical reality is a question best left aside (Tyrrell 1984: xiii). But regardless of their historicity, the Amazons certainly represented culturally and psychologically salient objects within the Greek mythological tradition, and the most obvious comment to make is that their society constituted an inversion of the established social order (Tyrrell 1984: 40ff.): women ruled and men were compelled 'like our married women'[25] to spend their time indoors carrying out the orders of their wives – that is, if they were not dispensed with altogether. As the speakers of Lysias' *Funeral Oration* declaimed:

> Long ago there were Amazons, daughters of Ares, who lived beside the river Thermodon. They were considered men on account of their courage rather than women on account of their physical nature, so much did they seem to excel men in their spirit rather than fall short of them because of their form.
>
> (Lysias 2, *Funeral Oration*: 4)

The idea of a society in which the normal pattern of sexual domination is reversed seems to hold widespread fascination and the notion of an Amazonian society has continued to captivate the imaginations of later ages. The well-known but unsubstantiated theories of 'primitive matriarchies' formulated in the nineteenth century and still occasionally cited today must owe something of their popularity to the appeal inherent in the idea of simply inverting the known social order in order to bring about what appears to be an obvious logical possibility.[26] Yet although in the Greek case Amazon society in many respects does represent an inversion of gender roles so that men become like women, and women like men, at the same time the Amazons can also be seen to represent the imaginative *extension* of certain ideas about women well established within male-dominated society. In other words. what was imagined was not simply a society in which women behaved like men but, rather, what a society would be like if it were *ruled* by women.

In this respect the habitat of the Amazons is the first point to consider. Though it is uncertain and shifting, it is at all times outside the Greek world. Indeed, as the Greek world expands, so the location of the Amazons recedes (Tyrrell 1984: 56). But that the Amazons live beyond the frontiers of the civilized world does not merely reflect the Greeks' failure actually to encounter them except in a remote and mythic past. As Tyrrell argues, what is significant about the Amazons' location is simply that it is outside the known world, beyond that frontier which divides civilization from savagery and culture from all that is opposed to culture. The Amazons are placed in a territory which is as much a conceptual space as a geographical space, and which is inhabited by both the subhuman and the suprahuman, the monstrous and the divine – centaurs, gorgons, cannibals on the one hand; the blessed Ethiopians who are closest to the gods, the inhabitants of Atlantis, or Odysseus' wondrous Paiakians on the other (Tyrrell 1984: 56–9). Amazon society, which could not be countenanced as part of civilization, is placed beyond it in a bizarre and barbaric realm.

In this context Aristotle's argument might be recalled: it is in Hellenic communities that the proper relationship between the sexes is to be found due to the existence of free men fit to rule, while among barbarian communities women are reduced to the status of slaves because in such communities everyone is without distinction a slave. The attributes of freedom, of order, of civilization are thus lacking among barbarians. *A fortiori* it could be argued that a society which even more grossly transgressed the requirements of civilization, not only by failing to have men fit to rule, but by actually allowing the subordination of men to women, would by definition have to be a barbarian society beyond civilization's bounds. But a slightly different construction could also be made of this logic: not only that a society ruled by women would have to be a barbarian society, but that one of the paradigms for a barbarian society would be one ruled by women – women possessed of those very qualities antithetical to order. In the description of contemporary barbarians – who were deemed to be wild, irrational, lacking self-control – it is not surprising to find them also characterized as effeminate.[27] In the imaginative construction of an ultimately barbarian society, it is not surprising to find it represented as a society of women. In short, the location of women's own society beyond the limits of the civilized world posits in extreme geographical terms the nature of women's tenuous incorporation into male society and their embodiment of those qualities which it had to displace.

It is important, then, to consider exactly what parts of the feminine stereotype the Amazons contradicted or rejected, for the rejection is by no means wholesale. As Tyrrell argues (1984: 52–5), it centres on the Amazons' refusal to marry and to accept any form of subordination to men. Either the Amazons mutilated their right breasts or they refused to suckle their infants; in either case what is denied is women's *domestic* role as mother and nurse – precisely the role celebrated in the *Thesmophoria* and in the worship of Demeter, which associated women's fecundity with civilization and the cultivation of crops. Moreover, in one account at least, the Amazons are ignorant of agriculture, which once more neatly combines the two aspects of fertility central to civilization.[28] The specificity of this rejection is significant, for it does not entail a denial of sexuality itself or of the commonplace attributes of feminity, but only of the domestic *function* of women's sexuality and of women's *social* role within civilized society.

The Amazons are not chaste; they do not do without men for their sexual gratification. Indeed, both in pictorial art and in literary accounts they are presented as figures of eroticism.[29] In fact they were promiscuous and according to one account, the male children whom they bore were simply distributed to the men with whom they had randomly mated since there was no way of knowing their particular paternity.[30] What is thus denied is what for the Greeks was a central fact of civilization: the ordered arrangement of sexual relations for the procreation of legitimate children to be models of their fathers. With the Amazons we may be in a bizarre and barbaric world, but at the same time we are not so far from the desires of women in Aristophanes' *Ekkleziazusai* or their cavalier attitude towards legitimacy in the *Thesmophoriazusai*. A society of women would know sex and it would know erotic desire, but it would not know marriage (cf. Tyrrell 1984: 53,66).

But if the Amazons' promiscuous eroticism removes them from the patronage of Demeter, pushing them outside culture and placing them in the province of Aphrodite, their rejection of marriage also situates them squarely in the realm of Artemis. The Amazons are, of course, no virgins; but, as we have seen, both Aphrodite and Artemis – excessive eroticism and excessive chastity – make common cause in their opposition to the civilized ideal of marriage and to the model of woman in her domesticated state. If the young virgin can be portrayed as a creature of the wild under Artemis' protection this is because, in the normal course of events, virginity is the status which precedes her marriage and domestication; but with the Amazons who reject marriage and are never domesticated, myth can present women in a permanent state of female savagery which includes sexual licence. And, as daughters of Ares the god of war,[31] the Amazons are devotees of an equally violent Artemis – of Artemis of Ephesus and of Thracian Artemis Tauropolos – of, indeed, that Artemis whom Nilsson felt to be at odds with the figure he derived from classical literature but who fuses in cult with Cybele the Phrygian Mother-goddess, for both Artemis of Ephesus and Artemis Tauropolos were goddesses whose essentially foreign rites were orgiastic (Tyrrell 1984: 86). Indeed, the Amazons were said to have founded the famous shrine of Artemis at Ephesus where, in myth, they took refuge on a number of occasions.[32] In Greece itself, in Lakonia, the temple of Artemis Astrateia, 'Artemis Ender of War', was said to be so named because the Amazons had stopped there from further warfare against Greece and dedicated a wooden statue of the goddess in the temple (Engle 1942: 532–4).[33] Moreover, it is worth noting that, as befits their association with Artemis, the Amazons are always portrayed as young women. There are no representations in Greek art of an elderly Amazon (Engle 1942: 517). The young maiden, a creature of the wild before marriage, is in Artemis' care; the Amazons, wild and permanently rejecting marriage, are the devotees of Artemis and always young.

But the Amazons' association with savagery and with a world of nature is conveyed not only by their devotion to Artemis. It is inherent in their mythological association with the Centaurs. Pomeroy has drawn attention to the fact that representations of battles between Greeks and Amazons ('Amazonomachies') are often paired, as on the Parthenon metopes, with representations of battles between Greeks and Centaurs. 'Perhaps the Greek mind, with its penchant for combining symmetry and alternatives, may have fictionalized the two groups, the Centaurs male and lustful, the Amazons female and chaste' (1975: 25). The juxtapositioning of Amazons and Centaurs is certainly interesting, and both Amazons and Centaurs also enter into the myths of Theseus who aids Peirithous against the Centaurs when they disrupt his wedding with Hippodamia,[34] while Peirithous in turn aids Theseus in an expedition against the Amazons.[35]

The nature of this mythicial *quid pro quo*, however, suggests that the pairing of Amazons and Centaurs in Greek art may have more to do with their similarities than with their differences. With the notable exception of the wise Cheiron, the Centaurs, half-man half-horse, were powerful figures of savagery – lewd, drunken, eaters of raw flesh, and in both form and characteristics placed somewhere between men and beasts.[36] It is in this state that they disrupt Peirithous' wedding and attempt to rape his bride. In form the Amazons were normal women (the possible mutilation of their right breast notwithstanding), but their portrayal as mounted horse-women makes them very much the Centaurs' counterparts – nor should it be forgotten that, contrary to our own cultural tradition, the horse itself was seen as a savage beast. Moreover there are indications that, like the Centaurs, the Amazons were at one stage in their tradition eaters of raw flesh,[37] and also that they dressed in animal skins[38] – features which again epitomize the bestial and the barbaric. As for their 'chastity', it amounts to no more than their refusal to enter into the domestic subordination of monogamous marriage; and here again they parallel the Centaurs, for the Centaurs did not have wives nor did they respect the laws of marriage (as their attempted rape of Hippodamia at her wedding celebration demonstrates). Centaurs and Amazons are parallels: savage, bellicose, uncivilized. But if the Centaurs are arrived at by combining man and beast, the Amazons are arrived at simply by postulating a society of women unruled by men.

In this context, the adventures of Herakles and Theseus are of considerable interest. The myths concerning both are numerous and complex in their constant reformulations and literary embroiderings, particularly so since the Athenians tried to promote Theseus into a national hero by assigning him feats and adventures closely modelled on Herakles (Kirk 1974: 109; cf. Connor 1970), for Herakles was one of the oldest heroes of Greek mythology and the most widely popular throughout Greece. Ridding the land of all manner of monsters, Herakles is as close to a 'culture hero' as can be found, and although parts of the Theseus story are obviously ancient, Theseus, 'owes much of his mythical *persona* to the desire of Athenians, and especially the tyrant Peisistratus in the sixth century BC, to make of him a great national hero. They did so in two ways: by associating him as closely as possible with Heracles, the *beau ideal*, and by ascribing to him various political and benevolent acts that were held to be the beginning of Athenian democracy' (Kirk 1974: 152; cf. Connor 1970).

Herakles' ninth labour was the task known as 'The Girdle of Hippolyte', the Amazon queen. Whether the girdle was an article of clothing or a piece of armour is uncertain, but Euripides mentions that it was 'preserved in Mycenae'.[39] At all events, Herakles obtained it, though there is some doubt as to whether he killed Hippolyte in the process (Kirk 1974: 187; cf. Tyrrell 1984: 2). In some versions Hippolyte visits Herakles on his arrival at the river Thermodon and offers him the girdle, but:

> Hera in the likeness of an Amazon went up and down the multitude saying that the strangers who had arrived [Herakles and his companions] were carrying off the queen. So the Amazons in arms charged on horseback down on the ship. But when Herakles saw them in arms, he suspected treachery and killing Hippolyte stripped her of her girdle. And after fighting the rest he sailed away.
>
> (Apollodoros, *Library*: 2.5.9)

In other versions, Herakles captures Hippolyte's sister, Melanippe, and ransoms her to Hippolyte for the price of the girdle.[40] (In some versions Theseus is made to accompany Herakles and is given another Amazon queen, Antiope, as his slave.)[41] As Kirk points out, however, the relatively numerous archaic vase-paintings which represent this episode of Herakles' life, and which carry it back to the end of the seventh century at least, do not show any girdle but represent an all-out battle between male and female warriors. They also tend to identify the Amazon queen with Andromeda or Andromache rather than Hippolyte (1974: 188). But whether or not the girdle was a purely classical invention, Herakles' subjection of the Amazons is an integral part of his mythic adventures and in one late account he destroyed both Gorgons and Amazons on his expedition to Africa since his resolve to be the benefactor of all mankind forbade letting 'any nation be under the rule of women' (Engle 1942: 534–5).[42]

As for Theseus, after his return from Crete and his adventures with the Minotaur, he is credited with the unification of Attica. Some ancient authorities make him the founder of constitutional government in Athens and the initiator of what was to become the democracy.[43] Certainly Theseus enjoyed a privileged position in the Athenians' own versions of their constitutional history,[44] and it is after the reforms which he brought about, setting government on a solid footing, that Athens' favourite son went on an expedition to the land of the Amazons. As already noted, some versions of the myth have Theseus accompanying Herakles, from whom he received Antiope as a share of the booty. In other versions Theseus launched his own expedition, accompanied by Peirithous. Antiope (or Hippolyte) then either falls in love with Theseus and follows him back to Athens, or is abducted by him and becomes variously his wife or his mistress, at all events the mother of his son, Hippolytos.[45] But the abduction/seduction of Antiope/Hippolyte results in the most interesting feature of the myth(s). The Amazons invade Attica, and Theseus is forced to engage in a protracted conflict with them which Athens barely succeeds in winning.[46]

It is, then, significant that both the great Greek culture hero, Herakles, and the Athenian national hero, Theseus, are made to confront and conquer the Amazons. In the Herakles myths the Amazons are classified with those other monstrous beings whom Herakles clears from the face of the earth, his expedition against the Amazons following from his capture of Diomedes' flesh-eating horses. And it is not difficult to understand why the Greeks should have conceived of a female-dominated society as an intolerable deviation against which their male hero, Herakles, had to pit himself in the process of tidying up the world.

Nevertheless it is still far from clear that the Amazons were seen as monstrous because they were 'contrary to nature'; rather, they were contrary to *society* and thus numbered among those phenomena of nature whose wildness civilization had to master. The Amazons represented women as they might well have been, freed from the constraints of male command, the ultimate extension of a view which saw the young virgin as a creature of the wild to be tamed and domesticated by marriage. Girls who were unbroken horses in the commonplace metaphors of Athenian society become in myth, and beyond the bounds of civilization, the wild horse-riding Amazons. Either Herakles, the Greek champion, breaks them by force (by killing Hippolyte and routing the Amazons) or Theseus exploits that weakness innate to all those who are not civilized and who are not, in the Greek sense, 'men' – their susceptibility to desire. But in either case there is a strong element of sexual dominance in the reduction of the Amazons to their required place. Herakles' theft of the 'girdle' amounts to a symbolic rape (Tyrrell 1984: 91), while in later versions of the Amazon myths, in which Theseus plays the hero, the theme of sexual dominance is even more pronounced. Hippolyte (or her sister Antiope) falls madly in love with Theseus and follows him of her own free will back to Athens where she becomes his mistress or wife. As the Athenenian orator Isokrates relates the story:

> our country (Attica) was invaded by the Thracians . . . and also by the Skythians led by Amazons, the off-spring of Ares, who made the expedition to recover Hippolyte, since she had not only broken the laws which were established among them, but had fallen in love with Theseus and followed him from her home to Athens, and there lived with him.
>
> (Isokrates 12 [*Panathenaikos*]: 193; Loeb trans., modified)

And so successful is Theseus' mastery of Antiope that in some versions she actually dies fighting at his side against the Amazons who have invaded Attica to avenge their honour and claim her back.[47] Again, it should be noted how this story recalls and combines standard themes: Kirke's change from hostility to devotion after Odysseus has slept with her; Andromache's claim that one night in bed dissolves a woman's hostility towards the man who has taken her; and, more generally, the repeated motif in which a man captures his bride from the world of nature to integrate her into the society to which he belongs. But perhaps the most interesting aspect of the myths of Theseus and the Amazons is their 'political' content.

It must be stressed that for the Athenians the Amazons' invasion of Attica was a matter of historical fact. The struggles against the Thracians and against the Skythians and Amazons, however mythical, stood out as the first great conflicts with barbarians and were classed with their triumph over the Persian invasion.[48] As Isokrates explains:

> Now, while the most celebrated of our wars was the one against the Persians, yet certainly our deeds of old offer no less strong evidence for those who dispute over ancestral rights. For while Hellas was still insignificant, our territory was invaded by the Thracians, led by Eumolpos, son of Poseidon, and by the Skythians, led by Amazons, daughters of Ares – not at the same time, but during the period when both races were trying to extend their dominion over Europe . . .
>
> In truth, they were not successful; no, in this conflict against our forefathers alone they were utterly overwhelmed as if they had fought the whole world. How great were the disasters which befell them is clear; for the tradition respecting them would not have persisted for so long a time if what was then done had not been without parallel. At any rate, we are told regarding the Amazons that of all who came not one returned again, while those who had remained at home were expelled from power because of the disaster here.
>
> (Isokrates 4 [*Panegyrikos*]: 68–70; Loeb trans., modified)

Inasmuch as a pseudo-chronological sequence of events can be constructed from the myths, it would thus seem that the very first threat which Theseus' newly federalized Attica, with its constitutional form of government and its new laws and institutions, had to face was an invasion of wild, barbarian horse-riding women from the north. Here, then, in remarkably clear-cut terms, is a confrontation between culture and nature, civilization and barbarity that was simultaneously a confrontation between the social and political achievements of men and the wildness and savagery of women. But Athens won, and in the orator Lysias' almost smug account (from which I have already quoted) the reasons for Athens' victory are clear. They relate to those innate qualities which, in the end, must entail women's subordination to society and to men:

> They were considered men on account of their courage rather than women on account of their physical nature, so much did they seem to excel men in their spirit rather than fall short of them because of their form. Ruling over many peoples, they had in fact enslaved those who lived near them, but by report they had heard of the renown of this country, and moved by the thought of great glory and by their great ambition they mustered the most warlike of their peoples and marched against this very city. But having met with good men, they found that their spirit [*psyche*] was now like their physique, and they achieved a reputation which was the reverse of their former one, and by their perils rather than by their bodies they appeared women. They alone failed to learn from their mistakes so as to be better advised in the future, nor would they return home to report their calamity and the valour of our forefathers. They died on the spot and paid the penalty for their stupidity, making this city's memory immortal because of its valour whilst they rendered their own native land nameless as a result of their disaster here. Thus by their unjust desire for others' land they justly lost their own.
>
> (Lysias 2 [*Funeral Oration*]: 4–6)

The Amazons, having encountered 'good men' – Greeks, Athenians, the forefathers of the present *polis* – are for the first time reduced to what they always were: mere women, motivated by greed, destroyed by folly, and incapable even of learning better for the future. As such the Amazons provide the prototype of all those who would challenge Athens' greatness and power, and challenge it from the peripheries of civilization. Politically, the reference is to the Persians, effeminate barbarians. Mythologically, their analogues are women, nature's barbarians.

The fusion of barbarity and women is not, however, limited to the creation of the Amazons. In many ways the theme is a constant one, but the location of women outside the boundaries of civilization is most clearly realized by two other representations, that of Dionysos' Bacchants and Medea the witch. These may most profitably be viewed in the context of Euripides' two tragedies, the *Bacchai* and the *Medea*.[49]

V

The god Dionysos' opening speech in the *Bacchai* is worth quoting at length:

> I am Dionysos, the son of Zeus,
> come back to Thebes . . .
>   Far behind me lie
> those golden-rivered lands, Lydia and Phrygia,
> where my journeying began. Overland I went,
> across the steppes of Persia where the sun strikes hotly
> down, through Bactrian fastness and the grim waste
> of Media. Thence to rich Arabia I came;
> and so, along all Asia's swarming littoral
> of towered cities where Greeks and foreign nations,
> mingling, live, my progress made. There
> I taught my dances to the feet of living men,
> establishing my mysteries and rites
> that I might be revealed on earth for what I am:
> a god.
>   And thence to Thebes.
>   This city, first
> in Hellas, now shrills and echoes to my women's cries,
> their ecstasy of joy
>
> . . .
>
> I have stung them with frenzy, hounded them from home
> up to the mountains where they wander, crazed of mind,
> and compelled to wear my orgies' livery.
> Every woman in Thebes – but the women only –
> I drove from home, mad. There they sit,
> rich and poor alike, even the daughters of Kadmos,
> beneath the silver firs on the roofless rocks.
> Like it or not, this city must learn its lesson:
> it lacks initiation in my mysteries
>
> . . .
>
>   On, my women,
> women who worship me, women whom I led
> out of Asia where Tmolus heaves its rampart
> over Lydia!
>   On, comrades of my progress here!
> Come, and with your native Phrygian drum –

Rhea's drum and mine – pound at the palace doors
of Pentheus!
(Euripides, *Bacchai*: 1–61; W. Arrowsmith trans, modified)

As the prologue makes clear, Dionysos' followers are in the first instance women: both the band which he led from Asia to be his companions and the women of Thebes whom he has driven in frenzy to the mountains. Admittedly Dionysos has no intention of confining his worship to women. The seer, Teiresias, and the old king, Kadmos, join the Bacchants and advise King Pentheus to do the same. Dionysos' express purpose is to make Pentheus recognize his divinity and to have his own religion established throughout Greece. It might be assumed that he was successful, for in the classical period Dionysos was honoured as a god by all, and the religious festival during which the *Bacchai* was performed was itself the festival of Dionysos with the god's image and his priests presiding. Euripides' last play thus recalls not only the purported origins of Dionysiac religion in Greece, but also the origins of the dramatic festival for which he had for so long written. But though established as a god among men, and as a god whose acceptance into Athens was officially celebrated by the ritual entry of his statue into the city, the fact remains that in myth it was women who first felt Dionysos' power.

For Pentheus, it is precisely Dionysos' effect on women that he finds most offensive – though he will not attribute their behaviour to divine cause. For Pentheus, the Theban women's worship of the new god is but an excuse for some drunken, libidinous orgy:

I happened to be away, out of the city,
but reports reached me of some strange mischief here
stories of our women leaving home to frisk
in mock ecstasies among the thickets on the mountain,
dancing in honour of the latest divinity,
a certain Dionysos, whoever he may be!
In their midst stand bowls brimming with wine.
And then, one by one, the women wander off
to hidden nooks where they serve the lusts of men.
Priestesses of Bacchos they claim they are,
but it's really Aphrodite they adore.

(ibid., 215–25)

Nor, so far as Pentheus is concerned, are women thus acting out of character. As he angrily declares to Teiresias and Kadmos, who are going to join the Bacchants:

> . . . When once you see
> the glint of wine shining at the feasts of women,
> then you may be sure the festival is rotten
> (ibid., 260–2)

Aristophanes, after all, dwells on a similar thought. And expressing sentiments remarkably similar to those of both Eteokles and Kreon, when they find themselves confronted by Thebes' women, Pentheus cries:

> . . . We march
> against the Bacchai! Affairs are out of hand
> when we tamely endure such conduct in our women
> (ibid., 784–6)

Dionysos (in disguise) may assure Pentheus that matters can be peaceably settled, but again the reply is:

> How?
> By accepting orders from my own slaves?
> (ibid., 803)

Dionysos' appearance, too, associates him with femininity – as Pentheus derisively notes when he orders Dionysos' capture:

> . . . go and scour the city
> for that effeminate stranger, the man who infects our women
> with this strange disease and pollutes our beds.
> (ibid., 352–4)

and again when he eventually interviews the disguised Dionysos:

> your *are* attractive, stranger, at least to women –
> which explains, I think, your presence here in Thebes.
> Your curls are long. You do not wrestle, I take it.
> And what fair skin you have – you must take care of it –
> no daylight complexion; no, it comes from the night
> when you hunt Aphrodite with your beauty.
> (ibid., 453–8)

Pentheus, of course, makes Dionysos' effeminate appearance a sign of womanizing lechery – a good, if complex, example of the tendency to blur physical sexual distinction by classing pretty boys and girls together, while maintaining a moral distinction by categorizing eroticism as a whole as effeminate and resistance to sexual desire as masculine. In Pentheus' eyes Dionysos is *both* effeminate *and* heterosexually lascivious, for both characteristics contrast with the austere ideal of the true man.

But the most powerful portrayal of the essentially feminine (or rather anti-masculine) nature of Dionysos' worship occurs in the magnificent scene in which Pentheus finally falls victim to Dionysos' powers. Dionysos (still in disguise) persuades Pentheus that he must go to the mountains to spy on his mother and the other Bacchants. To do so he must dress himself as a woman to avoid detection. At first Pentheus protests:

> *D*: First, however,
> you must dress yourself in women's clothes
> *P*: You want *me*, a man, to What?
> You want *me*, a man, to wear a woman's dress.
> But why?
>
> (ibid., 821–3)

But as Pentheus falls under Dionysos' hypnotic spell, his protests are mingled more and more with fascinated delight at the idea of transvestism:

> *P*: Dress? In a *woman*'s dress,
> you mean? I would die of shame.
> *D*: Very well.
> Then you no longer hanker to see the Mainads?
> *P*: What is this costume I must wear?
> *D*: On your head
> I shall set a wig with long curls.
> *P*: And then?
> *D*: Next, robes to your feet and a net for your hair.
> *P*: Yes? Go on.
> *D*: Then a thyrsos for your hand
> and a skin of dappled fawn.
> *P*: I could not bear it.
> I *cannot* bring myself to dress in women's clothes.
> *D*: Then you must fight the Bacchai. That means bloodshed.
>
> (ibid., 828–37)

Pentheus, now completely under Dionysos' control, exits to clothe himself in women's garb. Dionysos rejoices and lays bare his plot:

> Women, our prey now thrashes
> in the net we threw. He shall see the Bacchai
> and pay the price with death.
> O Dionysos
> now action rests with you. And you are near.
> Punish this man. But first distract his wits;
> bewilder him with madness. For sane of mind
> this man would never wear a woman's dress;
> but obsess his soul and he will not refuse.
> After those threats with which he was so fierce,
> I want him made the laughing stock of Thebes,
> paraded through the streets, a woman.
> Now
> I shall go and costume Pentheus in the clothes
> which he must wear to Hades when he dies, butchered
> by the hands of his mother. He shall come to know
> Dionysus, son of Zeus, consummate god,
> most terrible, and yet most gentle, to mankind.
> (ibid., 846–61)

A choral ode intervenes, and then the crazed Pentheus returns to suffer, unwittingly, Dionysos' mockery:

> *D*: Let us see you in your woman's dress,
> disguised in Mainad clothes so you may go and spy
> upon your mother and her company. Why,
> you look just like one of the daughters of Kadmos
> . . .
> *P*: Do I look like anyone?
> Like Ino, or my mother, Agave?
> *D*: So much alike
> I almost might be seeing one of them. But look:
> one of your curls has come loose from under the snood
> where I tucked it.
> *P*: It must have worked loose
> when I was dancing for joy and shaking my head.
> *D*: Then let me be your maid and tuck it back.
> Hold still

P: Arrange it. I'm in your hands.
completely.
D: And now your strap has slipped. Yes,
and your robe hangs askew at the ankles.
P: I think so.
At least on my right leg. But on the left the hem
lies straight.
D: You will think me the best of friends
when you see to your surprise how chaste the Bacchai are.
P: But to be a real Bacchante, should I hold
the wand in my right hand? Or this way?
D: No.
In your right hand. And raise
it as you raise your right foot. I commend your change of
heart.

(ibid., 914–44)

Thus Pentheus goes to his destruction.

Again the *Bacchai* is about much more than a conflict between men and women or between traditionally masculine and feminine roles. Perhaps it is about the necessity of accommodating within society the irrational, ecstatic, supposedly divinely inspired aspects of existence; about the folly of refusing to recognize divinity in its more savage forms; about the over-evaluation of human order and the presumptuous elevation of human reason to the extent that they are thought proof against that which they cannot comprehend; about man's failure to recognize his mortal place.[50] As Teiresias warned Pentheus near the play's beginning:

> Do not be so certain that power
> is what matters in the life of man; do not mistake
> for wisdom the fantasies of your sick mind.
> Welcome the god to Thebes; crown your head;
> pour him libations and join his revels.
>
> (ibid., 310–13)

But Euripides allows no easy moral to be drawn: Dionysos' final epiphany is quite uncompromising in the punishments it brings.[51] And if Kadmos can plead, horrified, that:

> Gods should be exempt from human passions
>
> (ibid., 1348)

Dionysos' cold reply is simply that:

> Long ago my father, Zeus, ordained these things.
> (ibid., 1349)

In Euripides' profoundly pessimistic view man may learn by suffering, but there is no forgiveness and he usually suffers. But if the *Bacchai* is about more than a conflict between men and women, traditional notions of gender differentiation are still called into play. Doubtless we are meant to understand that Pentheus was wrong in denying Dionysos' divinity and in seeing his effect merely in terms of the riotous behaviour his worship causes among women – just as Kreon was wrong in refusing to bury Polyneikes and in reducing the importance of his burial to a trial of strength with an insubordinate female. In both cases what women argue for is divinely sanctioned and, in the final analysis, is seen as socially and morally correct. But despite the validity of their stance, the behaviour of the women in the *Bacchai* still draws on a conception of women as the natural enemies of order. That Pentheus, a man, should stand as the champion of social propriety, rationality, scepticism (albeit in a narrow and reprehensible form) while Dionysos and his followers – effeminate and female – should stand as the representatives of the unrestrained, the wild, the irrational, the ecstatic (albeit in a divinely sanctioned form) is no more accidental than that the person who should challenge Kreon's laws should be an Antigone, or that those expressing fear and panic in the face of Eteokles' stern ordinances should be Thebes' women. If the moral of the *Bacchai* lies in the necessity of society and of men to welcome the wild and the irrational in the form of Dionysos' divinity, it is nevertheless the case that the wild and the irrational already have their latent presence in society in the form of women. And Dionysos does not simply destroy Pentheus as Thebes' king; he quite literally destroys his mind and with that his masculinity. Rightly or wrongly, the threat to society's stability and to the rationality upon which it is based comes regularly from women.

But consonant with the essentially 'feminine' nature of Dionysiac religion, it should also be noted that, despite Dionysos' birth from a Theban princess, Dionysos, his followers, and his religion are all presented as essentially non-Greek. Dionysos has travelled from Lydia, leading his band of women through Phrygia, Persia, Bactria, the land of the Medes, Arabia, and the Asiatic coast of the Aegean.

Dionysos himself associates his religion with Cybele and Rhea – female and ecstatic Asiatic deities. With somewhat alarming naivety, Nilsson saw the 'victorious march of Dionysus' as a genuine historical movement which 'spread in the form of a violent physical epidemic, almost like St Vitus' dance, more particularly amongst women, since they are specially susceptible to this kind of infection (sic). . . . The traces of the severe struggle by which Dionysus prepared the way for the ecstatic cult have been converted into myths' (1964: 206–8). But it is no more necessary to consider the historicity of Dionysos' Asiatic origins and his 'introduction' into Greece than to consider the historicity of the Amazons to register the essential fact that the Greeks saw Dionysos and his followers as 'foreign'. That is significant in itself. As with the Amazons, myth places Dionysos and his ecstatic women – women who have broken free from the place assigned them by society – among the barbarians. It places them geographically as well as morally outside the accepted order of civilization. And while with the Amazons it is a barbarian tribe of women that has to be repulsed by the newly formed Attic state, with Dionysos and his maenads it is a barbarian religion infecting women that temporarily returns them to their natural savagery.

But not only does the Asiatic religion which captivates the Theban women come from outside the ordered world of the Greek *polis*; Dionysos literally drives the women from the city to the mountains where, inspired by the god, they comport themselves in a manner reminiscent of Artemis and her followers.

The brilliant messenger speech describes the scene in all its variety:

> our grazing herds of cows had just begun to climb
> the path along the mountain ridge. Suddenly
> I saw three companies of dancing women,
> one led by Autonoë, the second captained
> by your mother Agave, while Ino led the third.
> There they lay in the deep sleep of exhaustion,
> some resting on boughs of fir, others sleeping
> where they fell, here and there among the oak leaves –
> but all modestly and soberly, not, as you think,
> drunk with wine, nor wandering, led astray
> by the music of the flute, to hunt their Aphrodite

through the woods.
But your mother heard the lowing
of our hornèd herds, and springing to her feet,
gave a great cry to waken them from sleep.
And they too, rubbing the bloom of soft sleep
from their eyes, rose up lightly and straight –
a lovely sight to see: all as one,
the old women and the young and the unmarried girls.
First they let their hair fall loose, down
over their shoulders, and those whose straps had slipped
fastened their skins of fawn with writhing snakes
that licked their cheeks. Breasts swollen with milk,
new mothers who had left their babies behind at home
nestled gazelles and young wolves in their arms,
suckling them. Then they crowned their hair with leaves,
ivy and oak and flowering bryony. One woman
struck her thyrsus against a rock and a fountain
of cool water came bubbling up. Another drove
her fennel in the ground, and where it struck the earth,
at the touch of god, a spring of wine poured out.
Those who wanted milk scratched at the soil
with bare fingers and the white milk came welling up.
Pure honey spurted, streaming, from their wands.
(ibid., 677–711)

The vision is idyllic. The women are free, a bountiful world of nature magically at their command. Wolf-cubs and fawns suckle at their breasts; snakes lick their cheeks; under Dionysos' spell they have all become, like Kirke, *potniai theron*, 'mistresses of nature'. Now it is not only virgins who are 'unyoked', but all women, young and old.

Dionysos the foreign god and his barbarian religion have returned women to the natural state from which men and civilization plucked them. At the approach of men, however, the scene changes and the women's savagery is unleashed:

but she [Agave] gave a cry: 'Hounds who run with me,
men are hunting us down! Follow, follow me!
Use your wands for weapons'.
At this we fled
and barely missed being torn to pieces by the women.

> Unarmed, they swooped down upon the herds of cattle
> grazing there on the green of the meadow. And then
> you could have seen a single woman with bare hands
> tear a fat calf, still bellowing with fright,
> in two, while others clawed the heifers to pieces.
> There were ribs and cloven hooves scattered everywhere,
> and scraps smeared with blood hung from the fir trees.
> And bulls, their raging fury gathered in their horns,
> lowered their heads to charge, then fell, stumbling
> to the earth, pulled down by hordes of women
> and stripped of flesh and skin more quickly, sire,
> than you could blink your royal eyes. Then,
> carried up by their own speed, they flew like birds
> across the spreading fields along Asopos' stream
> where most of all the ground is good for harvesting.
> Like invaders they swooped on Hysiai
> and on Erythrai in the foothills of Kithairon.
> Everything in sight they pillaged and destroyed.
>
> . . .
>
> The Bacchai then returned where they had started,
> by the springs the god had made, and washed their hands
> while the snakes licked away the drops of blood
> that dabbled their cheeks.
>
> (ibid., 731–68)

The mainads change from gentle nymph-like creatures to frenzied monsters as nature goes on the rampage. It is thus that Pentheus the man, the city's ruler, will meet his death, torn to pieces by the Bacchantes led by his own mother who, still delirious, will proudly enter Thebes exhibiting the bloody head of her son, mistaken for a lion. The play thus exploits in its description both a vision of women as innocent creatures of the wild about to be surprised and captured – in this case by Pentheus' herdsmen who would carry them back to Thebes – and a vision in which they are the embodiment of savagery, attacking and destroying all they meet. On the one hand they are gentle nurses giving suck to fawns and wolf-cubs; on the other wild beasts who tear cattle to shreds and rip King Pentheus limb from limb.

Like Artemis and her nymphs they are part of an idyllic world of nature from which men would abduct them; but like the wilder aspects of Artemis – like Atalanta – they are themselves dangerous huntresses (cf. Segal 1984: 206); dressed in skins they are eaters of raw flesh and – like the Amazons – they attack the settlements of men. And although Euripides seems at some pains to stress that the women of Thebes are not indulging in sexual licence (or drunken debauchery as Pentheus assumes), still the Chorus sings:

> O let me come to Cyprus,
> island of Aphrodite,
> homes of the loves that cast
> their spells on the hearts of men!
>
> . . .
>
> Bromios, take me there!
> There the lovely Graces go,
> and there Desire, and there
> the right is mine to worship
> as I please.
>
> (ibid., 402–15)

Moreover, whatever their actions, they are performed under the sway of a god whose essence lies in the relaxation of rational self-control and an abandonment to those mysterious forces which lie beyond the rational self. The play may preach acceptance of Dionysos and celebrate his incorporation into the city, but it still leaves little doubt that what must be accepted is a frightening and dangerous thing. Pentheus (like Kreon) is destroyed by what he cannot comprehend, and the women, representatives of all which lies outside the city's limits, not only overthrow its ordered rule, but destroy the very basis of its structure. Agave kills not only Pentheus the king, but Pentheus her son. State and family perish together.

In dramatizing what purports to be an historical and singular event – the introduction of Dionysos' worship into Greece – and in dealing with a subject whose import certainly extends beyond the respective characteristics of men and women, Euripides is still drawing on a series of established themes which express in a diffuse manner the relationships between men and women and women and society. The traditional connection between Dionysos and women can itself be seen as the manifestation of an underlying if never entirely explicit formulation of female nature. The formulation derives from the constant association of a series of oppositions between culture and nature, civilization and barbarity, the rational and the irrational, self-control and emotionality, which finally accrete themselves around the opposition between male and female.

When, rather than dwelling confidently on the ordered relationship between men and gods celebrated in the institutions of civic worship, Greek religious thought chose instead to reflect uneasily on the mysterious facets of divinity that challenged society and emphasized the precarious nature of civilization's achievements, it saw in women the occasional harbingers of divine retribution (cf. Segal 1984: 195–6). Women were indispensable to society and yet they were never fully part of it; they represented within society something of that which lay outside it (cf. Gould 1980: 57–8). And if what lies beyond the compass of society and the rational control of men tends to merge (if for no other reason than its incomprehensibility) with the divine, then it was through women that those forces which society had failed to take into account might make their inroads (cf. Gould 1980: 58). Indeed, it appears regularly to have been women who became, in the strict sense of the word, enthused, who mediated between gods and men. It was, for example, Kassandra in the grip of Apollo – a very different god from Dionysos – who saw in her mantic trance the fall of Troy and the ancient horrors of the House of Atreus into which she had been led a captive; who saw against her will and by divine grace the workings of that relentless curse which would destroy the best laid plans of Greece's most successful king (again, at the hands of a woman). But if such a view of women appears a little lofty, it might also be remembered that those same qualities which could draw women into association with the divine were usually construed in less elevated fashion. Women might be the natural registers of Dionysos' mysterious divinity, but their affinity with him could also be seen at a more basic level – in those plays of Aristophanes, for example, where, rather than being enthused with the god, woman is simply a drunken sot; rather than yearning for the divine, she is rushing into the arms of a lover; rather than transcending man's rationality, she is simply incapable of grasping it. Those same qualities which, in the exceptional circumstances of myth might make her appear close to divinity, could equally justify the guarded control under which society placed her.

VI

If the Bacchai are 'normal' Greek women turned – or returned – by Dionysos' divinity to a state of nature (and to their 'natural' state), by contrast Medea, the 'mythical archetype of a sorceress' (Kirk 1974:237), is, on the face of it, one of the most bizarre figures of the mythological tradition.

The main outlines of Medea's history may be summarized as follows.[52] Daughter of Aietes, King of Kolchis, Medea was granddaughter of Helios the sun and the niece of both Pasiphai and Kirke.[53] When the Greek hero Jason sailed with the Argonauts to Kolchis to regain the Golden Fleece, Medea fell in love with him. On condition that Jason marry her and take her back to Greece, Medea aided him with her magic unguents – proof against fire and metal – to fulfil the tasks set by her father: to yoke the two fire-breathing brazen-footed bulls, plough the field of Ares, and sow it with dragon's teeth left over from Kadmos' sowing of Thebes.[54] But, the tasks completed, Aietes revoked his bargain. Medea again came to Jason's aid. She cast a spell on the dragon that guarded the Fleece, allowing Jason to steal it while the dragon slept. Accompanied by Medea and her half-brother, Apsyrtos, Jason and his crew set sail. Aietes pursued them, but on Medea's advice Apsyrtos was dismembered and thrown overboard piece by piece, thus delaying Aietes who collected the remnants of his son for burial.[55]

After further adventures Jason and Medea arrived in Greece. Jason was heir to the throne of Iolkos. However his uncle, Pelias, had overthrown Jason's father, Aison, and killed him during Jason's absence.[56] Medea offered to dispose of Pelias. She visited Pelias, claiming she could rejuvenate him – a feat she demonstrated by dismembering a ram, boiling it in a cauldron, and extracting a lamb. Having charmed Pelias to sleep, she ordered his daughters similarly to dismember their father and place him in the same cauldron. Pelias, however, remained in pieces. For this crime both Jason and Medea were exiled, and made their way to Corinth where they lived for ten years.[57]

In Corinth, Jason wished to divorce Medea to marry Glauke, daughter of King Kreon. Enraged, Medea sent Glauke a magnificent robe which consumed her as soon as she put it on, in turn killing Kreon as he embraced his dying daughter. At this point there is considerable divergence. According to some accounts, Medea left her children as suppliants at the altar of Hera where they were seized and killed by the Corinthians – for which crime the Corinthians thenceforth performed annual expiations.[58] According to the version followed by Euripides, it was Medea who killed her children by Jason in further revenge. Medea then fled Corinth in a marvellous chariot drawn by winged serpents sent to her by her grandfather Helios, the sun.

Medea then sought sanctuary with King Aigeus of Athens whom she married. When Aigeus' son, none other than Theseus, arrived in Athens, Medea plotted against him. Aigeus was unaware of Theseus' paternity, and being persuaded by Medea that Theseus was a traitor, sent his son to fight the Marathonian bull. Theseus returned successful, and so Medea attempted to poison him; the plot failed, Theseus was recognized by Aigeus, and Medea was again expelled.[59] According to some accounts, Medea then returned to Kolchis where she restored her father to the throne after he had been deposed by his brother Perses and finally ended her days in the Elysian Fields as Achilles' consort.[60]

Medea, then, was no ordinary woman. Nor, of course, is she portrayed as Greek – her homeland is Kolchis, mysterious and distant. But she was certainly a figure of Greek mythology, and within that context it remains to be seen to what extent she is a truly 'alien' being or to what extent, like the Amazons and the Bacchantes, her characteristics are consonant with that stereotypical vision of women to which an alien quality is in any case integral.

Let us consider the question of sorcery. Medea is the possessor of occult knowledge of medicines and drugs. She protects Jason with her unguents against the fire-breathing brazen-footed bulls; charms the sleepless dragon to sleep; rejuvenates the ram (and deceptively proposes to rejuvenate Pelias). She also heals the wounds of the Argonauts, and it was her promise to King Aigeus of Athens (who thought himself childless) that she could ensure him a son by her potions that persuaded him to give her sanctuary.[61]

Equally, Medea is a poisoner. She slays Glauke and Kreon with a poisoned robe and attempts to poison Theseus. There is not, however, much difference between Medea's poisons and her magic potions: in Greek both are simply *pharmaka*, 'drugs'. Nor, as a woman, was Medea alone in their use. To mention but a few examples from myth, Pasiphai bewitched her husband, Minos, so that whenever he took another woman to his bed he discharged snakes, scorpions and millepedes from his joints – a disadvantage which was dealt with by a further magic potion when Prokris desired to sleep with him.[62] Kirke too administered a drug to Odysseus' men to make them forgetful of their homeland and she tried to poison Odysseus.[63] Kreousa attempted to poison Ion and Xouthos with Gorgon's blood – one drop of which healed disease, the other of which caused death. And Deianeira accidently killed her husband Herakles in exactly the manner in which Medea disposed of Glauke, with a robe soaked in poison which ate away his flesh. Rather like Kreousa's poison, it was distilled from a Centaur's blood, though poor Deianeira believed it to be a love potion.[64]

Nor is it only in myth that women were represented employing such devices. As Ehrenberg observes, 'Women were supposed to be favourite customers of those who traded in drugs, particularly love potions' (1974: 198), and Asiatic women had the reputation of being experts in their use (152). Equally, the portrayal of women as poisoners occurs in Aristophanes, and if the law-court speech Antiphon 1 [*Stepmother*] is a genuine prosecution, then we seem to have a case of life imitating art; for there the confusion between love potions and poisons is remarkably similar to Deianeira's in the *Trachiniai*. As the speaker, the defendant's stepson, relates:

> In the first place, I was ready to torture the defendant's slaves, who knew that this woman, my opponent's mother, had planned to poison our father on a previous occasion as well, that our father had caught her in the act, and that she had admitted everything – except that it was not to kill him, but to restore his love that she alleged herself to be giving the potion.
> (Antiphon 1 [*Stepmother*]: 9; Loeb trans., modified)

Apparently undeterred, the speaker's stepmother continued with her plot, this time successfully and with the aid of an innocent slave-girl, the mistress of one of her husband's friends:

> There was an upper room in our house occupied during his visits to Athens by Philoneos, a highly respected man and a friend of our father's. Now Philoneos has a *pallake* whom he proposed to place in a brothel. My brother's mother made friends with her; and on hearing of the wrong intended her by Philoneos, she sends for her, informing her on her arrival that

> she was also being wronged by our father. If the other would do as she was told, she said, she herself knew how to restore Philoneos' love for her and our father's for herself.
>
> . . .
>
> Later, Philoneos happened to have a sacrifice to perform to Zeus of the Household.. . . After supper was over, the two [Philoneos and the speaker's father] set about pouring libations . . . But Philoneos' *pallake*, who poured the wine for the libation while they offered their prayers – prayers never to be answered, gentlemen – poured in the poison with it. Thinking it a happy inspiration, she gave Philoneos the larger draught; she imagined perhaps that if she gave him more, Philoneos would love her the more: for only when the crime was done did she see that my stepmother had tricked her. . . . So they poured their libation, and, grasping their own slayer, drained their last drink.
>
> (ibid., 14–19)

The confusion between love potions and poisons in the above cases occurs within one of the most common employments of magic – to ensnare a lover. Indeed it should be noted that the drugs and poisons which Medea used, and which Kreousa, Pasiphai, and Prokris used, though not in themselves love potions are all nevertheless employed in an erotic context: Medea's love for Jason, Kreousa's jealousy and fear of being discarded, Pasiphai's desire to keep Minos faithful, and Prokris' desire to sleep with him. As it happens, Medea herself was the victim of a form of erotic magic employed by Aphrodite and specifically associated with women. As Detienne explains, a certain bird, the *iunx*, or wry-neck, was attached by Aphrodite to a wheel. With its four members bound in a circle which turned endlessly it became the powerful charm which attracted Medea to Jason. The *iunx* was, in Pindar's words, the 'delirious' bird.[65] In fact, the term *iunx* had three separate but closely related denotations: the wry-neck, a bird famed for its incessant movement; a female magician expert in erotic magic; and an erotic charm. This last *iunx* was a small wheel pierced with two central holes through which chords were threaded. By pulling the ends of the two chords, the wheel spun, emitting a whistling noise. And in the poem by Theokritos entitled *The Sorceresses*,[66] nine incantatory couplets are pronounced by a woman separated from her lover, each followed by the refrain, '*Iunx*, draw to my dwelling this man, my lover' (Detienne 1977: 83–6). Again one is reminded of Kirke, Kalypso and the Sirens.

Medea may then be the mythical archetype of the sorceress, but her knowledge and use of drugs, poisons, and magic made her less a peculiarity than the embodiment of certain proclivities attributed to women in general. Or, to put matters the other way, it is significant that those who had the skills of the magician in Greek mythology, and those who were deemed to make use of magic and poison in practice, were routinely women. The use of magic and poison was a feminine art. But though poisons and magic might have been female tools, their association with women can also be seen to have an aptness beyond any consideration of empirical fact. Within the Greek tradition the forces of the erotic – of the emotions, the passions, and physical desires – were considered as external agencies, as *nosoi*, sicknesses, which could destroy men. By the same token, magic charms, potions, and poisons appear as but a concretization of those feminine emotional drives already considered external forces. The whirling stone, the delirious bird, the *iunx*, representing in physical form the frenzied madness of *eros* and women, driven by their own unchecked passions, were the obvious candidates to employ such devices against men. Medea is the sorceress and poisoner *par excellence*; but equally her motivations are the standard female jealousies and desires.

Here the nature of Medea's crimes should be recalled. She betrays her father Aietes. She kills her half-brother, Apsyrtos. She beguiles Pelias' daughters into slaying their own father. She murders Jason's bride and father-in-law. She kills her own children. She attempts to poison Aigeus' son, Theseus. Motivated by passion, aided by magic, Medea thus lays waste every household within which she is situated and every society with which she has contact. In the person of Medea, those forces of eroticism and desire which had to be suppressed or controlled within civilized life now poison the fabric of society. And although Medea is clearly portrayed as a foreigner, her barbarian origins – like those of the Amazons and the mainads – relate her once again to a Greek conception of women as outsiders brought into a civilization to which they do not truly belong and to which they introduce those dangerous emotional forces also foreign to civilized society.

In Euripides' play the *Medea*, however, Medea's traditional characteristics are maintained while at the same time being used once again to question the values of civilization and the ambitions of men. Euripides' play deals with the period of Medea's domicile with Jason in Corinth, with Jason's attempt to marry the Corinthian princess Glauke, and with Medea's revenge. Again Medea is portrayed as a foreigner, a homeless transient insecure in her Corinthian refuge. When she has murdered Glauke, Kreon, and her own children to appear before Jason in the serpent-drawn chariot of her grandfather Helios, Jason's horrified cry is:

> There is no Greek woman who would ever have dared such deeds.

Jason's immediate recourse is to classify Medea as something totally foreign to Greek society:

> Out of all those whom I passed over and chose you
> To marry instead, a bitter destructive match,
> A monster, not a woman, having a nature
> More savage than Tyrrhenian Skylla.
> (Euripides, *Medea*: 1340–3; R. Warner trans, modified)

However, as Shaw (1975) has cogently argued, far from appearing as an alien monster whose actions could be explained away on these grounds, Medea is presented as typically feminine and typically Greek – as, indeed, are the moral values she espouses. 'Inside the house the bonds between friends are permanent, and they are of the blood. In the case of man and wife these bonds are children, and Medea tells Jason she could forgive him if he were still childless' (Shaw 1975: 259). Medea's reactions to Jason's conduct are thus exceptional in degree, not in kind. Her impassioned speech to the chorus of Corinthian women is an acute summary of the Greek woman's condition (as Euripides saw it) and not the expression of some barbaric mentality:

> But on me this thing has fallen so unexpectedly,
> It has broken my heart. I am finished. I let go
> All my life's joy. My friends, I only want to die.
> It was everything to me to think well of one man,
> And he, my own husband, has turned out wholly vile.
> Of all things which are living and can form a judgement
> We women are the most unfortunate creatures.

Firstly, with an excess of wealth it is required
For us to buy a husband and take for our bodies
A master; for not to take one is even worse.
And now the question is serious whether we take
A good or bad one; for there is no easy escape
For a woman, nor can she say no to her marriage.
She arrives among new modes of behaviour and manners,
And needs prophetic power, unless she has learned at home,
How best to manage him who shares the bed with her.
And if we work out all this well and carefully,
And the husband lives with us and lightly bears his yoke,
Then life is enviable. If not, I'd rather die.
A man, when he's tired of the company in his home,
Goes out of the house and puts an end to his boredom
And turns to a friend or companion of his own age.
But we are forced to keep our eyes on one alone.
What they say of us is that we have a peaceful time
Living at home, while they do the fighting in war.
How wrong they are! I would very much rather stand
Three times in the front of battle than bear one child.

(ibid., 225–51)

In fact *both* Jason and Medea are foreigners in Corinth. Both are cut off from their own society. It is precisely Jason's attempt to integrate himself into Corinthian society by marrying its princess which causes Medea to commit the crimes for which, later, he berates her in horror. Jason acts out of social and political expediency, and that expediency means the rupture of his relationship with Medea. Medea acts in defence of her marriage to Jason – in defence of a solemn compact which, in her view, cannot expediently be dissolved. It is because she will not see herself and her children cast aside that, perhaps paradoxically, she prefers to destroy herself and them. Jason's claim that Medea is an unnatural and alien monster contains an element of obvious hypocrisy, or at least comes with a certain irony. His own actions, which caused Medea's crimes against family and society, are in themselves a betrayal of the values of domestic life; but between Jason, the man, and Medea, the woman, there is also no consensus about what those values entail.

Medea, as she makes clear on a number of occasions, has relinquished everything for Jason:

> Where am I to go? To my father's?
> Him I betrayed and his land when I came with you.
> To Pelias' wretched daughters? What a fine welcome
> They would prepare for me who murdered their father!
> For this is my position – hated by my friends
> At home, I have, in kindness to you, made enemies
> Of others whom there was no need to have injured.
> And how happy among Greek women you have made me
> On your side for all this! A distinguished husband
> I have – for breaking promises. When in misery
> I am cast out of the land and go into exile,
> Quite without friends and all alone with my children.
> (ibid., 502–13)

Her commitment is to Jason alone, and Jason has betrayed her and her trust. He has disregarded everything upon which she has built her life. But Jason's world is a different one. For him what is important is not private devotion but public honour, esteem, and position (Shaw 1975: 260–1). As he reasons with Medea:

> When I arrived here from the land of Iolkos,
> Involved, as I was, in every kind of difficulty,
> What luckier chance could I have come across than this,
> An exile to marry the daughter of the king?
> It was not – the point that seems to upset you – that I
> Grew tired of your bed and felt the need of a new bride;
> Nor with any wish to outdo your number of children.
> We have enough already. I am quite content.
> But – this was the main reason – that we might live well,
> And not be short of anything. I know that all
> A man's friends leave him stone-cold if he becomes poor.
> Also that I might bring my children up worthily
> Of my position, and, by producing more of them
> To be brothers of yours, we would draw the families
> Together and all be happy. You need no children.
> And it pays me to do good to those I have now
> By having others. Do you think this a bad plan?
> You wouldn't if the love question hadn't upset you.
> (ibid., 551–68)

Clearly the family is as important to Jason as it is to Medea. In

Jason's own opinion it is more important to him than to Medea. He even asks why Medea has need of children. But Jason and Medea are working with different conceptions of the family. For Medea it is the domain of intimate relationships, to be valued in themselves, an enclosed and self-sufficient world. For Jason it is part of the structure of larger society, the means by which he must find his place in the public world. For Medea, marriage means her exclusive attachment to a man; for Jason, marriage means his manifold attachment to society. Jason's and Medea's values are completely at odds even if they both centre on the same thing. In the opposition between Medea and Jason we are presented with an extreme contrast between feminine and masculine viewpoints, between the emotional values of personal ties and the practical values of social position. Indeed, Jason refers everything to its publicly recognized worth. By citing the benefits of living in the civilized Greek society to which he has brought her, Jason can even argue that Medea has already profited sufficiently from her marriage to him (Shaw 1975: 259–61).

But Jason's expressions of esteem for the values of Greek society – for law, justice, fame – are surely meant to sound hollow. On hearing of Medea's treatment by Jason and her plans for revenge, the chorus of Corinthian women have already sung:

> Flow backward to your source, sacred rivers,
> And let the world's great order be reversed.
> It is the thoughts of *men* that are deceitful,
> *Their* pledges that are loose.
> Story shall now turn my condition to a fair one,
> Women are paid their due.
> No more shall evil-sounding fame be theirs.
> (ibid., 410–20)

Any feelings aroused that Jason is being portrayed as a callous opportunist with a gift for rationalization are endorsed by this ode. There can be little doubt that the sympathies Euripides creates are meant to be with Medea in particular and women in general. The commonplace contrasts between men and women, rationality and emotionality, public and private are being used in order to point towards significant moral inadequacies in the values of his time. The dialogue between Medea and Jason does more than underscore characteristic differences between feminine and masculine viewpoints. It acts as an indictment of those who would justify self-interest on the grounds of rationality. Jason's justifications for his actions are not in accordance with justice; they are a perversion of it. Once again it is women who are presented as in touch with an instinctive morality that is devoid of sophistry.

But if it is by taking the woman's point of view that Euripides shows up Jason's 'rationality' as self-seeking calculation, there is still no suggestion that virtue consists in the untempered instincts. Society must still be guided by reason. The emotions to which women are subject remain dangerous forces. When Jason, echoing the words of Hippolytos, declares that:

> But you women have got into such a state of mind
> that, if your life at night is good, you think you have
> Everything; but, if in that quarter things go wrong,
> You will consider your best and truest interests
> Most hateful. It would have been better far for men
> To have got their children in some other way, and women
> Not to have existed. Then life would have been good.
>
> (ibid., 569–75)

We might well be inclined to think he is mouthing misogynist platitudes. They are certainly not in accord with the sympathy Euripides shows towards women. And yet, ultimately, the play bears out the truth of Jason's words. Jason is careful to stress that his marriage to Glauke is not prompted by erotic desire; he *must* marry to become a member of Corinthian society – and it is this marriage which proves his undoing. Were women not involved, Jason's social aims could have been achieved without conflict (cf. Shaw 1975: 261). Yet because social ambitions of necessity involve women, immediately those violent passions are unleashed which destroy his plans. It is still women who are subject to the emotions, and it is still the emotions which result in chaos and strife.

Here it should be noted that, for all the sympathy Euripides seems to create for Medea, the version of her story he relates is bloodier than most. For it is not, as in other accounts, the Corinthians who kill her children after she has murdered Glauke and Kreon, but Medea herself.[67] Her revenge is unreasoning, excessive, deranged. And when the chorus of Corinthian women, who have thus far endorsed Medea's plans, learn of her full intent, even they sing in terror. Medea's own words are the most revealing:

> Come, children, give
> Me your hands, give your mother your hands to kiss them.
> Oh the dear hands, and O how dear are these lips to me,
> And the generous eyes and the bearing of my children!
> I wish you happiness, but not here in this world.
> What is here your father took. Oh how good to hold you!
> How delicate the skin, how sweet the breath of children!
> Go, go! I am no longer able, no longer
> To look upon you. I am overcome by sorrow.
> I know indeed what evil I intend to do,
> But stronger than all my rational thought is my *thymos*
> That which, brings upon mortals the greatest evils.
> (ibid., 1069–80)

Medea is in anguish. She is not some alien monster. She is a Greek mother. But her *thymos*, her feelings, her passions, her emotions – however we care to translate the term, this non-rational faculty of the *psyche* – overcomes or conquers any rational appreciation of the situation. Her *thymos* is placed in clear opposition to her intellect. But Medea's *thymos* is *kreisson*, 'stronger, mightier'. As Aristotle commented, in woman the rational faculty of the *psyche* is *akyron*, 'incapable', 'not-in-command'. Medea will kill her own children; but as she is also made to say, the *thymos* is the cause of man's greatest evils.

It is from this point that Medea becomes truly barbaric; or rather, it is from this point that the play starts to emphasize the horror of her deeds, her derangement, and the complete contravention of civilized life which her actions entail.[68] The messenger appears on stage and describes in detail the deaths of Glauke and Kreon. Medea rushes out to slay her children with a sword. We hear their pathetic cries. The chorus sings:

> O Earth, and the far shining
> Ray of the Sun, look down, look down upon
> This poor lost woman, look, before she raises
> The hand of murder against her flesh and blood.
> Yours was the golden birth from which
> She sprang, and now I fear divine

Blood may be shed by men.
O heavenly light, hold back her hand,
Check her, and drive from out the house
The bloody Fury raised by friends of Hell.
(ibid., 1251–61)

But Euripides creates a final surprise, a last twist to the play's meaning. As Jason appears fresh from the scene of Glauke and Kreon's deaths to accost Medea and to learn that his own children too have been slain, Medea, in a magnificent *coup de théâtre*, appears above their house in the chariot of Helios. Far from smiting Medea down (as the chorus prayed, and as Jason too thought would be fitting) Helios has come to rescue her. Medea triumphs and miraculously departs, leaving Jason and Corinth shattered behind her. At the very time we see Medea at her most barbaric, now a truly alien being standing in the wondrous chariot of the sun, we are reminded not only of her affinity with those powerful natural forces which lie outside the bounds of the civilization, but also of the fact that her actions have been sanctioned by the gods.

As Shaw (1975: 283) remarks, Medea is now essentially where Jason found her. She has returned to that savage, barbaric, strange – but also semi-divine – state from which Jason took her. But inasmuch as Medea has also been made in so many respects to represent the typical Greek woman, we must also see even Euripides' complex rendition of the myth touching on that theme whereby women in general are viewed as alien to society and as beings who bring into its midst those dangerous forces antipathetic to its order. Jason's crime was to disregard Medea's love; but his folly was not to foresee the madness that could cause. It is, then, of some importance to note the chorus that extols the virtues of the Athens to which Medea will flee:

From of old the children of Erechtheus are
Splendid, the sons of blessed gods. They dwell
In Athens' holy and unconquered land,
Where famous Wisdom feeds them and they pass gaily
Always through that most brilliant air where once, they say,
That golden Harmony gave birth to the nine
Pure Muses of Pieria.

And beside the sweet flow of Kephisos' stream,

Where Kypris [Aphrodite] sailed, they say, to draw the water,
And mild soft breezes breathed along her path,
And on her hair were flung the sweet-smelling garlands
Of flowers of roses by the Loves [*Erotes*], the companions
Of Wisdom [*Sophia*], her escort, the helpers of men
In every kind of excellence.

(ibid., 824–45)

Athenian self-congratulation, perhaps; but the point made is that in Athens, fairest and most civilized of cities, the powers of Aphrodite and of Eros are not totally rejected. Moreover, they are coupled with wisdom, *sophia*, in a harmony whereby they contribute to virtue. And as the chorus have already sung:

When the Loves [*Erotes*] come in excess
They bring men no honour
Nor any worthiness.
But if in moderation Kypris [Aphrodite] comes,
There is no other power at all so gracious.
O goddess, never on me let loose the unerring
Shaft of your bow in the poison of desire.

Let temperance [*sophrosyne*] protect me,
It is the gods' best gift
On me let dread Kypris

Inflict no wordy wars or restless anger
To urge my passion [*thymos*] to a different bed.
But with discernment may she guide women's weddings,
Honouring most what is peaceful in the bed.

(ibid., 627–42)

Such is the ideal state – one of temperance and moderation. Confined and regulated within marriage, those affections to which women are subject should unite a family; rejected, or unrestrained, they will tear society apart. As Shaw writes, 'Medea's poison . . . is not magic but a symbol. It represents the ability to create, destroy, or pervert the bonds between father and child. It is love, and its god (Hekate) lives in the hearth. It is fire, and its user descended from the sun' (1975: 260). Euripides would give the passions of women their due; but the message remains that with women and the passions one is still playing with fire.

VII

In discussing the Amazons, the *Bacchai*, and the *Medea*, and in looking at Artemis and Aphrodite, I have, in what is necessarily a brief and partial account of some aspects of Greek mythology, concentrated on those female representations which are perhaps extreme in the manner in which they locate women, both geographically and morally, outside the bounds of society. I have also omitted much of the mythical and ritual material which, by contrast, tends to stress women's integration into society. But though the examples may be extreme, they do not, I think, yield a distorted picture of the Athenians' conception of women – or at least of a conception inherent in their literary, mythical, and ritual traditions. Sophokles' Deianeira may be, for example, a much more accurate representation of the submissive Athenian wife than Euripides' Medea; but Deianeira is still 'Man-destroyer', and it is still her passion for Herakles that causes his death, and a wild Centaur who wreaks his vengeance on Greece's culture-hero through the medium of a woman. Euripides' *Iphigeneia at Aulis* may present us with a young heroine prepared to die for her father and her fatherland, but it is still Artemis, goddess of the wild, who rescues her and who, in *Iphigeneia in Tauris*, transports her to the barbaric land of Thrace where she becomes the sacrificer of men. And in Aischylos' retelling of the myth of the Trojan War, Helen, so magnificently dwelt on by the chorus of the *Agamemnon*, becomes the outsider embraced by both the house of Atreus and of Priam to their mutual destruction; she becomes desire incarnate, a bewitching phantasm driving men to insanity; she becomes the lion-cub brought up in the house, the bird whom the luckless boy cannot hold. Aischylos' is a very sophisticated rendition, but the imagery with which he supplements the myth repeats the constant theme – the fusion of the female, the wild, the beautiful and the destructive.

Again, such rituals as the *Thesmophoria* may express the sacred participation of women in the continuing life of the community, but what is celebrated is still society's harnessing of and coexistence with those forces of nature and of the divine which lie beyond its bounds and to which women belonged. Hesiod's story of Prometheus and the creation of Pandora, which is articulated with the themes of the loss of primal bliss and the instigation of a new order, shows both the necessity of woman to society and her estrangement from its aims. It illustrates the paradox whereby woman is the consumer of men, their sex, their strength, their food, and their wealth, and the instigator of all the evils in the world; yet without her society cannot continue, for the golden age is over and man, no longer part of nature, must work it and women to exist.

A question remains, however. In the traditions of myth and ritual, women were part of nature rather than culture. Consonant with those traditions, women were characterized in Athenian writings as emotional and unrestrained rather than rational and self-controlled. Consonant with that characterization, women were excluded from the secular life of the *polis* and relegated to the domestic sphere. Is it not then odd that Athens' eponymous patron, symbol of her civic order, and indeed goddess of wisdom, should be Athena, a female deity?

First, it should be emphasized that the connection I have tried to show between the mythical role of women, the literary representations of women's natural characteristics, and the position of women in the social organization of the Athenian state, is not intended to prove that myth, 'psychological' characterizations and social structure were all part of an indissoluble unity. We are dealing with an *historical* society – that is, a society whose various institutions, traditions, beliefs, and notions must be seen as having their own and to an extent autonomous developments. Athena was one of the oldest members of the Greek pantheon. Her existence as Athens' patron goddess has nothing to do with women's position within the social structure of classical Athens or with its notions about the nature of women. Her identity with Athens is as old as Athens itself. It was an historical fact for the Athenians as much as it is for us, and an immutable one. But inasmuch as this goddess came to stand for a variety of qualities which were considered non-feminine – wisdom, sense, reflection, military prowess, the creation of arts and crafts – we should also note what came to be her most salient characteristic: her total asexuality. She is Athena Parthenos, Athena the virgin; but unlike Artemis' virginity, Athena's is not a-marital, but androgynous. Blessed with a goddess as patron deity, Athens turns her into a transvestite, sexless, cerebral creature who springs fully armed and fully formed from the head of Zeus. As Kirk comments, 'there is a smack of sophistication and scholarship about this piece of theogony' (1974: 121). Indeed there is; for if Athens' patron deity had perforce to stand for those qualities –

masculine qualities – which were its civic ideals, then it was by such sophistication and scholarship that she renounced all ties with her sex. Athena is no longer a woman; she is the progeny of the mind of Zeus.

But perhaps this was not always enough. Perhaps in the context of fifth-century Athens and its male-dominated society the patronage of a goddess still seemed a little queer. In Aristophanes' *Birds*, Euelpides asks which god should be the protector of their new republic, Cloudcuckooland:

> – Why not retain Athena as City-keeper?
>
> – And how can that be a well-ordered State, where a goddess, born a woman, stands full-armed, and Kleisthenes [a notorious effeminate] holds a spindle?
>
> (Aristophanes, *Birds*: 828–31)

NOTES

1 INTRODUCTION: THE 'PROBLEM' OF WOMEN

[1] Belief and the problem of women was originally published in J. S. La Fontaine (ed.), 1972, *The Interpretation of Ritual*, but appears reprinted in S. Ardener (ed.) (1975) *Perceiving Women*. Page references are to the reprint. Cf. Gould (1980: 38–9) who also discusses Ardener's article.

[2] References to the conventional view opposed by Gomme may be found cited in his article (1925), but the views of, say, Tucker, who remarked that it was 'a great blot on Athenian civilization that the position of women had retrograded since the days of Homer' (1907: 51), or Wright, who went so far as to say that 'the Greek world perished from one main cause, a low ideal of womanhood and a degradation of women which found its expression both in literature and in social life' (1923: 1 cited in Pomeroy 1975: 6) are not unrepresentative. Cf. Lowes Dickinson, 'Nothing more profoundly distinguishes the Hellenic from the modern view of life than the estimate in which women were held by the Greeks' (1938: 70; first published 1896).

[3] A brief history of the controversy with references to a selection of the opinions expressed by classical scholars may be found in Richter (1971).

[4] Originally published in *Arethusa* 6 (1973), the bibliography is reprinted (with a supplement covering the years 1973–81) in Peradotto and Sullivan (eds) (1984) *Women in the Ancient World*. The *Arethusa Papers*.

[5] Gomme was no doubt correct in arguing against an 'Attic contempt' for women, but when for example, he states that '. . . for the rest, I consider it very doubtful if Greek theory and practice differed fundamentally from the average, say, prevailing in mediaeval and modern Europe' (1925: 25) and presents this as part of his claim that in Athens there was an 'equality between the sexes' (22), then one begins to realize the assumptions upon which his views are based.

[6] It is worth noting that Kitto continually invoked the relativity of cultural values in his assessment of the position of women in Athens, but again a discussion which starts with the words 'most men are interested in women, and most women in themselves' (1951: 219)

must place his own views within a particular historical and social context.

[7] It is instructive to note Peradotto and Sullivan's comments on the differences between the contributions to the 1973 *Arethusa* issue on 'Women in antiquity' and the issue which appeared five years later in 1978 (1984: 4–5).

[8] A full listing and discussion of Thoukydides' mentions of women will be found in Harvey (1985).

[9] For somewhat different approach to this question, see King (1983: 109–10).

2 POLITICS

[1] There is no specific statement in the ancient sources that women were forbidden to attend the assembly, but it is clear from Artistophanes' fantasy, the *Ekklesiazusai*, that they did not.

[2] Aristotle, *Politics*: 1278a 36.

[3] M. I. Finley, *The Ancient Economy* (1973: 48), estimates that the ratio of non-citizens to citizens ranged at different times from 1:6 to perhaps 1:2.5. The number of slaves in Athens seems almost anyone's guess, and estimates range from 20,000 to 400,000. In his most recent studies of Athenian demography, Hansen argues for a citizen population (adult males) of 60,000 in 432/1 BC declining to a minimum of 25,000 at the end of the century (1988: 26–7), and for an average ratio of citizens to metics in the fourth century of at least 3:1 (1988: 11). He also convincingly argues that the number of slaves in Athens will never be known since the Athenians themselves did not know, but for the sake of argument he places their number at approximately half that of the free population (1988: 11–12).

[4] For the penalty of slavery, see H.R.W. Harrison, *The Law of Athens Vol. I, The Family and Property* (1968: 165) and *Vol. II Procedure* (1971: 169).

[5] See Finley (1973). Even this disability was not so great, for non-citizens could lease agricultural land and rent domestic property. The privilege of owning land could also be conferred by special decree of the assembly (Harrison 1968: 199, 237–8). For full discussion of the political implications of land-ownership, see Peçirka (1963).

[6] I am conscious, in the above, of making rather large claims in a short space; but to substantiate them fully would be beyond the scope of this book. Though it is now outdated, Fustel de Coulange's *La Cité Antique* (1864) conveys better than any later work the closeness of the connection between the family and the state and the nature of their religious unity. Cf. Humphreys and Momigliano, 'The social structure of the city' in S. C. Humphreys (1978) *Anthropology and the Greeks*, first published in *Annali della Scuola Normale superiore di Pisa, ser. III*.4 (1974). Needless to say, when referring to the Athenian *polis* as bound together by ties of kinship, I am not suggesting that everyone was literally related to each other by consanguinity or affinity. It is a

question of ideology whereby cohesion is expressed through the *idiom* of kinship – by reference, for example, to a common ancestry.

[7] Aristotle, *Constitution of the Athenians*: 55.3. Cf. Harrison (1971: 202).

[8] See Aristophanes, *Acharnians*: 523ff. where Aspasia is blamed for the start of the Peloponnesian War; cf. Plutarch's *Life of Perikles*: 24.6 for mention of other comic references.

[9] For the use of *aste* and the restricted definition of *politis* see Gould (1980: 46 and notes 17 and 56).

[10] Aischines 1 Timarchos: 110–11.

[11] See, for example, T.G. Tucker, *Life in Ancient Athens* (1907: 51) or Jaeger (1939: 20–2). Both also believe that, here was something 'racial' or 'Asiatic' in the Athenians' treatment of women.

[12] Cf. Marylin B. Arthur, 'The origins of the western attitude towards women' (1984: 36). While believing that women's social role and function did not undergo any fundamental transformation, Arthur argues that the Greek city-state 'gave women status as an aspect of the men's existence, rather than as existants in their own right'.

[13] For a concise account of women's religious roles, see Pomeroy (1975: 75–8).

3 LEGAL CAPABILITIES

[1] Schaps' own conclusion is more detailed, and his important article 'The woman least mentioned: etiquette and women's names' (1977) demonstrates that there were in fact three categories of women whose names were freely mentioned in court: disreputable women, women connected with the speaker's opponent, and dead women. Given the general tactics of Athenian law-court oratory, the first two categories can almost be assimilated. The quite respectful mention of dead women's names might be explained by the fact that the passage of time would have made their identification by reference to male relatives somewhat difficult.

[2] Aristophanes, *Ekklesiazusai*: 1024–5. The women, who have taken over the state, now advert to the fact that no male *kyrios* can engage in contract over the value of one *medimnos* of barley. For further discussion of a woman's capability to engage in commercial transactions, see Schaps' excellent account (1979: 52–58).

[3] To take some examples from the speeches of Isaios: in Isaios 3 [*Pyrrhos*], Pyrrhos' estate was claimed by Phile, Pyrrhos' alleged daughter, who was represented by her husband, Xenokles. The estate was also claimed by Pyrrhos' sister, who was represented by her son. In Isaios 5 [*Dikaiogenes*], a share of Dikaiogenes' estate was claimed by his three surviving sisters, represented by Polyartos, the husband of the oldest sister, and then later by the sons of the sisters. In Isaios 7 [*Apollodoros*], the estate of Apollodoros was claimed by Apollodoros' first cousin, the daughter of Eupolis, represented by her husband Pronapes. In Isaios 11 [*Hagnias*], the estate of Hagnias was claimed by Philomache, the wife of Sosistheos, represented by her husband, and perhaps by

the mother of Hagnias, who was also her own son's second cousin.

For this last extremely complicated case (or series of cases) which should be read in conjunction with Demosthenes 43 [*Makartatos*], see the outstanding commentary of Wyse (1904). The mother of Hagnias was said to be represented by *kyrioi* – probably her sons. Most commentators also believe that in Isaios 10 [*Aristarchos*] the speaker was claiming the estate of Aristarchos as his mother's *kyrios* and not in his own name (Wyse 1904).

4 At least in Antiphon I [*Prosecution of the Stepmother*], a woman is prosecuted by her stepson for the murder of his father. She is, however, defended by her own sons, the prosecutor's homopatric half-brothers. (See also Harrison 1971: 84 and note 4.)

5 According to the speaker of Isaios 10 [*Aristarchos*], Aristarchos had two sons and two daughters, of which one of the daughters was the speaker's mother. One of the sons, Kyronides, was adopted by his maternal grandfather to become his grandfather's heir, and thus by the rules of Attic succession renounced all rights of inheritance to the estate of his father, Aristarchos. Thus, when Aristarchos died, his other son, Demochares, became sole heir to his father's property (for daughters had no rights of inheritance *if* there were sons). However, Demochares himself died while still a minor. Aristarchos' other daughter also died. Thus, with one brother dead, the other brother, Kyronides, having renounced his rights of inheritance, and her sister dead, the speaker's mother now stood to inherit the entirety of her father's property. However, her situation also meant that she was *epikleros*, i.e. with-the-property, and claimable in marriage, along with her property, by her father's closest male relative. Her closest male relative was her father's brother, Aristomenes, who was also, of course, her *kyrios*. He did not marry her himself, but instead married her off to an outsider, keeping the property she had inherited and bestowing it upon his own daughter whom he then married to none other than Kyronides. In essence, then, the speaker's claim is that his mother, who ought to have been an heiress and the sole inheritor of the property of her father, Aristarchos, was defrauded by a family conspiracy between her uncle (Aristarchos' brother) and her brother, Kyronides, both of whom ought to have been her protectors. A similar case of family conspiracy may also lie behind Isaios 3 [*Pyrrhos*]. For discussion of both, see Wyse's commentary (1904). For the rules of Attic succession and the situation of the *epikleros*, pp. 95ff.

6 Demosthenes 27 [*Aphobos I*] and Demosthenes 28 [*Aphobos II*].

7 E.g. Demosthenes 27 [*Aphobos I*]: 40; Demosthenes 29 [*Aphobos III*]: 26.

8 E.g. Demosthenes 41 [*Spoudias*]: 24; cf. Lysias 32 [*Diogeiton*]: 18 ff.

4 MARRIAGE AND THE STATE

1 See Schaps (1979: 41, 127): litigants appealed for sympathy on the grounds that they needed money to dower their marriageable daughters

(Demosthenes 40 [*Boiotos II*]: 4, 56, cf. Demosthenes 59 [*Neaira*]: 8), and attacked opponents by accusing them of causing women to remain unmarried (Demosthenes 59 [*Neaira*]: 112–13). Demosthenes 45 [*Eratosthenes*]: 21 includes among the crimes of the Thirty Tyrants that they caused women to remain unmarried. Cf. Lysistrata's complaints about the effects of the Peloponnesian War (*Lysistrata*: 529–30).

2 The only case in which the giving of a dowry was a legal requirement was when a girl whose father was a member of the *thetic* class became an *epikleros* – but without, of course, any property to go with her. Her next-of-kin was then obliged either to marry her himself or to dower her according to his means. The regulations are quoted in Demosthenes 43 [*Makartatos*]: 54. The rule is implied in Isaios 1 [*Kleonymos*]: 39, and perhaps in Demosthenes 59 [*Neaira*]: 113. See Harrison (1968: 135–6), Schaps (1979: 37–8). For a full discussion of the moral necessity of dowering women, see Schaps (1979: 77–81).

3 See Benveniste (1969: 239–44). There is no common Indo-European term for the institution of marriage. Cf. Vernant (1982: 49).

4 See also Gernet (1921: 339–43) and (Wyse 1904: 285–7). For cases of more distant relatives giving a girl in marriage by *engue*, see Lysias 32 [*Diogeiton*]. It appears that Diogeiton gave his brother's daughter in marriage (a girl who was simultaneously his daughter's daughter), but most probably his *kyrieia* over her was the result of his brother's will. Cf. Isaios 6 [*Philoktemon*]: 51 where it is stated that a girl, if she was not an *epikleros*, could be given in marriage by her collateral relatives to whom ever they chose.

5 Demosthenes 36 [*Phormion*]: 8, 28–30; Demosthenes 30 [*Onetor* II]; Demosthenes 57 [*Euboulides*]: 41; Plutarch [*Perikles*]: 24.

6 Demosthenes 27 [*Aphobos* I]: 5; Demosthenes 28 [*Aphobos II*]: 15–16; Demosthenes 29 [*Aphobos III*]: 43.

7 This is certainly the case in Demosthenes 57 [*Euboulides*]: 41. Although it was the husband who arranged the new marriage for his wife it seems that her brother acted as *kyrios* in actually giving the woman to her new husband. In Demosthenes 30 [*Onetor II*] the facts are very unclear. It could have been the husband, Timokrates, who gave his wife to Aphobos, or it could have been her brother, Onetor, who gave her. Most likely it was the latter, but obviously with the husband's consent. Perikles may also have given his wife in marriage with the consent of her former *kyrios*. At all events, if a woman's original *kyrios* (at the least her father) retained the right to dissolve her marriage, it would be only consistent that his permission would be needed for her husband to arrange any further marriage.

8 Wolff's brilliant article of 1944 remains the most convincing account of the legal rationale of Athenian marriage, but should be read in conjunction with Harrison (1968).

9 See Demosthenes 41 [*Spoudias*]: 6, Demosthenes 47 [*Euboulides*]: 41 and Isaios 8 [*Kiron*]: 14 for witnesses, and Isaios 3 [*Pyrrhos*]: 28–9 for the

argument that if there were no witnesses, then a marriage could not have taken place.

[10] In Lysias 19 [*Aristophanes*]: 14–16 the speaker's father took his mother without a dowry because of her father's virtue and fame. Wealthy young men were also willing to take the speaker's sisters without a dowry. In Demosthenes 40 [*Boiotos II*]: 25, the speaker says it would have been unfitting for his mother to have been married without a dowry. The inference is that she could have been married without one. See Harrison (1968: 49 and n. 1).

[11] See Isaios 3 [*Pyrrhos*]: 8–9, 28–9, 35–9, 78. Cf. Harrison (1968: 49 n.1).

[12] See Demosthenes 28 [*Aphobos II*]: 15ff. and Demosthenes 29 [*Aphobos III*]: 43.

[13] Demosthenes' mother remained a widow (Demosthenes 29 [*Aphobos III*]: 26) and Aphobos himself married the daughter of Philonides of Melite (Demosthenes 27 [*Aphobos I*]: 56). Demosthenes' sister was eventually married to her first cousin (mother's sister's son). See Davies (1971: 123, 142).

[14] See Isaios 6 [*Philoktemon*]: 47 for a further statement of the law.

[15] But see Humphreys (1974: 89 n.5) who argues that Aristophanes combined phrases from two laws, one concerning *nothoi* and one concerning intestate succession.

[16] It is clear from Antiphon I [*Stepmother*]: 14 that the term *pallake* could be used even of a slave woman whose master kept her for sexual pleasure.

[17] Plutarch, *Solon* 22.4.

[18] Such children were still, however, classified as *nothoi* (Wolff 1944: 86).

[19] Possibly these were women who had little or no dowry, or perhaps those whose chastity was under some suspicion, and who could not, consequently, be placed in an *oikos* as a legitimate wife – but this is speculation.

[20] References for the dispute over the civic status of *nothoi* before 1904 may be found in Wyse (1904: 278). See also Wolff (1944: 76 n.157). The dispute continues. See Humphreys (1974), MacDowell (1976) and Rhodes (1978), and, most recently, Hansen (1985).

[21] See Demosthenes 23 [*Aristokrates*]: 65, Demosthenes 59 [*Neaira*]: 104, Antiphon 5 [*Herodes*]: 62.

[22] It is worth noting that this exclusion is in no way denied by the evidence of Demosthenes 40 [*Boiotos II*]: 10 where it is suggested that two allegedly illegitimate men should be adopted by their uncles (mother's brothers) to relieve their alleged father of the necessity of introducing them into his phratry. The uncles might still swear to the men's legitimacy as a question of fact (Wolff 1944: 80–1; cf. Humphreys 1974: 89). Isaios 6 [*Philoktemon*]: 17ff. has sometimes been adduced as evidence that *nothoi* could be legitimized by admission to their father's phratry, but see Wyse (1904: 504–5). For discussion of Isaios 7 [*Apollodoros*]: 16–17: (quoted above) and of the impossibility of adopting *nothoi*, see Wyse (1904: 558–9).

[23] There is one case, Isaios 3 [*Pyrrhos*], in which it appears that an

illegitimate girl, a *nothe*, was given in marriage by *engue* to an Athenian citizen; by analogy it has been argued that a *nothos* could enter into marriage by *engue*. However, in Isaios 3 it is precisely the legitimacy of the girl, Phile, which is in question. See Wyse (1904: 278–9).

24 Notably by Wyse (1904: 278ff.), Wolff (1944) and Humphreys (1974). Harrison (1968: 61–8) does not believe that *nothoi* were excluded from citizenship. This is one instance in which I would depart from his authority.

25 See, for example, the argument of Ferguson (1910: 257–84) who believes that phratries became unimportant after Kleisthenes' reforms.

26 Much also hinges on the meaning of *eleutheros* (free) used by Aristotle in his account of qualifications for enrolment on the deme list. As Wyse (1904: 281) points out, '*eleutheros*' might well have meant 'of citizen birth' in this passage.

27 A fundamental question might be asked. Why, if citizenship was not dependent on legitimacy, did the state take the trouble to specify that no illegitimate could be a member of his father's *oikos*? The state would have had no interest in interfering with family structure unless family structure was directly articulated with the structure of the *polis*.

28 This is evident from Pollux's definition of a *nothos* – 'he who is born of a foreign woman or a *pallake*', and of a *gnesios* – 'he who is born of a citizen woman (*aste*) who is married (*gamete*)' (3.21).

29 For other mentions of Perikles' law, see Plutarch, Perikles: 6.10, 37.3, Suidas s.v. *demopoietos*.

30 I find Wolff difficult to follow at this stage. While maintaining that the statute 'was directed against an alien *hetaira* cohabiting with a citizen' (hence *synoikein* is being used in the broad sense of 'cohabitation'), he then goes on to argue that *synoikein* 'must be interpreted as a highly technical term if we are to understand the enactment'.

31 Such views, anachronistic though they may be, are worth bearing in mind if only because they have held such a prominent place in later western (Christianized) culture, and especially in the Mediterranean. The anthropological literature on the subject is extensive, but see, for example du Boulay (1986).

32 Again this is worth mentioning since although neither the Catholic nor the Orthodox churches have ever forbidden the remarriage of widows, in rural Mediterranean society (and especially in rural Greece) there has been a marked social barrier to the remarriage of widows precisely on the grounds that they are no longer virginal and no longer 'pure'.

33 See Plutarch, *Solon*: 23.

34 There is no actual mention of Solon's law being repealed, but it appears to have fallen into disuse. How a woman could be punished for adultery other than by her exclusion from membership of an *oikos* and from the religious life of the *polis* (unless the death penalty were invoked – of which there is no mention) is not easy to see. Her standard disabilities meant that she could not be fined, and she had no active civic rights which she could lose. On the other hand, it should be noted

that her exclusion could involve public humiliation. Thus in Aischines 1 [*Timarchos*]: 183 Solon's legislation is futher referred to:

> For the woman who is taken in adultery, he [*Solon*] does not allow to adorn herself, nor even to attend the public religious rites, lest by mingling with innocent women she corrupt them. But if she does attend them or adorn herself, he commands that any man who meets her shall tear off her garments, strip off her ornaments, and beat her (only he may not kill or maim her); for the law-giver seeks to disgrace such a woman and to make her life unlivable.
>
> (Loeb trans., modified)

35 For example Isaios 3 [*Pyrrhos*]: 29.

36 For details concerning 'dotal' mortgages, see Fine (1951: 116–41) and Finley (1951: 44–52).

37 For examples, see Demosthenes 59 [*Neaira*]: 52, Demosthenes 27 [*Aphobos I*]: 17, Isaios 3 [*Pyrrhos*]: 78.

5 FAMILY AND PROPERTY

1 See Demosthenes 57 [*Euboulides*]: 20 and Plutarch, *Themistokles*: 32. See Harrison (1968: 22) for further comment.

2 Nepos, *Kimon* 1.2, Philo Judaeus, *De specialibus legibus*: 3.4, scholia to Aristophanes, *Clouds*: 1371. See Harrison (1968: 22 and n.3).

3 Pomeroy (1975: 68–70) attempts some demographic speculation, and the possibility of female infanticide should not be underrated (cf. Lacey 1968: 164–7). Pomeroy (1983) examines the question of female infanticide in more detail, but the evidence relates primarily to the Hellenistic period.

4 See, for example, Demosthenes 59 [*Neaira*]: 8, Isaios 3 [*Pyrrhos*]: 8–10.

5 For a full account of the *anchisteia* see Harrison (1968: 143–9).

6 For the problems of interpretation related to Demosthenes 43 [*Markatatos*]: 51 and Isaios 11 [*Hagnias*]: 1–2 concerning the meaning of 'sons of cousins', see Wyse (1904: 671ff.) and Harrison (1947). See also Davies (1971: 77–89) for the prosopography.

7 See, for example, Isaios 2 [*Menekles*]: 15, Isaios 6 [*Philoktemon*]: 5, Demosthenes 43 [Makartatos] 84.

8 Exactly what the speaker is referring to by this last statement is not clear – probably no more than that it was the archon's responsibility to ensure the correct transmission of a man's *oikos* (Wyse 1904: 576).

9 E.g. Isaios 2 [*Menekles*]: 13; Isaios 3 [Pyrrhos]: 68; Isaios 4 [*Nikostratos*]: 16; Isaios 6 [*Philoktemon*]: 9; Demosthenes 20 [*Leptines*]: 102; Demosthenes 44 [*Leochanes*]: 68; Demosthenes 46 [*Stephanos II*] 14. See Gernet (1955: 121 and n.5).

10 Gernet (1955: 129–30) lists known cases of adoption and argues conclusively that although Athenians might be free to adopt whomsoever they wished, in fact they adopted close relatives.

11 The subject of *epikleroi* is dealt with in some detail by Harrison (1968: 9–12, 132–8), and Schaps (1979: 25–47). Perhaps the most

insightful work on the subject, however, remains that of Gernet (1921). The following account is based largely on the works of these three authors. Invaluable commentary is also supplied by Wyse (1904: *passim*).

12 See Gernet (1921: 353); the evidence is not, however, good.

13 The interpretation I have put forward largely follows Gernet (1921), but it must be admitted that it raises one of the most vexed issues concerning *epikleroi* and the whole purpose of the institution of the 'epiclerate', for it seems that in the fourth century the son of an *epikleros* was made his maternal grandfather's heir by the process of posthumous adoption. Schaps (1979: 32–3) thus argues strongly that in Athens the purpose of the epiclerate had nothing to do with the preservation of a man's *oikos* via his daughter. I believe, however, that Gernet's classic articles (1921) sufficiently account for Schaps' objections. See also Wyse (1904: 361–2). Admittedly, however, a number of factual questions about the institution of epiclerate remain unanswered for want of decisive evidence, and these bedevil the interpretation of its purpose. See Harrison (1968: 132–8) for a concise and authoritative summary.

14 See Wyse (1904: 672ff.) for the complicated history of this family and Davies (1971: 77–89).

15 For the complexities and uncertainties of this case, see Wyse (1904: 483ff.).

6 FREEDOM AND SECLUSION

1 Aristophanes: 549ff. See Ehrenberg (1943: 151 and n.8).

2 Aristophanes, *Acharnians*: 478, *Knights*: 19ff., *Thesmophoriazusai*: 387ff., *Frogs*: 840ff.

3 For an account of the Athenian *agora* and its inhabitants, see Ehrenberg (1974a: 113ff.). For an analysis of the social status of women involved in trade, see Herfst (1922: 77ff.). For an account of women's participation in trade, see Schaps (1979: 61–3). I have not touched on the intriguing problem of what was meant by the *agora gynaikeia* (see Herfst 1922: 36ff.). See also Lacey (1968: 170–2).

4 Euripides, *Elektra*: 309.

5 Menander, *Dyskolos*: 198ff.

6 Plutarch, *Perikles*: 10.4–5.

7 In *Gorgias*: 502 b-e Plato says that poetry in general and tragedy in particular are the kind of rhetoric addressed to 'boys, women and men, slaves and free citizens, without distinction'. In *Laws*: 817 a-c Plato says that in the ideal state there would be a reluctance to allow the tragic poets to 'erect their stages in the market-place and perform before women and children and the general public'. In *Laws*: 658 a-d Plato claims that if the audience were called upon to state their preferences, children would vote for the conjuror, boys for the comic poet, young men and the more refined sort of women for the tragic

poet. See Haigh (1907: 324–29), and for further evidence of women's attendance.

[8] See Haigh (1907: 327). The reference to miscarriages is in the *Life of Aischylos* (*Aischylou Bios*) which is included in the Oxford Classical Text edition of Aischylos.

[9] Dover (1972: 17) suggests that women may have congregated separately from their menfolk, seeing as much of the plays as they were able to after the men had taken their seats.

[10] See Aristophanes, *Lysistrata*: 641ff. For a brief account of women's participation in religious cult, see Pomeroy (1975: 75–8).

[11] See Demosthenes 59 [*Neaira*] 21 – though as it happens the woman is a slave. The Eleusynian Mysteries were open to all: male and female, free and slave.

[12] All Athenian women knew how to work wool, and this was a major domestic occupation. The story of Aristarchos and his womenfolk in Xenophon's *Memorabilia* 2.7.2–14 is, however, very interesting, for it was only in a time of crisis, and on Sokrates' radical advice, that their skills were put to commercial advantage. Once again, Aristarchos is a gentleman, a member of the leisured class; a considerable social inhibition had therefore to be overcome – but no such inhibition may have existed for the poorer citizens about whom we know less. For discussion of this passage, see Lacey (1968: 170–1).

[13] The anthropological literature that bears on this topic is now immense but see, for example, Dubisch (ed.) (1986) and Ardener (ed.) (1981). See also Gould's examples (1980: 48–9): one from Friedl (1962: 12) and one from personal experience in Crete. See also Hirschon (1978) cited in Gould (1980).

[14] Harvey (1984) has recently drawn attention to a much neglected fact: that Ischomachos' ideal wife was also one and the same person as the Chrysilla who, in Andokides' *On the Mysteries*, is alleged to have later become the mistress of her own daughter's husband, Kallias. This must modify our opinion of Ischomachos' wife herself; it does not, however, alter the import of the picture of ideal married life presented by Xenophon.

[15] For an excellent discussion of women in New Comedy and of the relationship between their romantic plots and Athenian social (and legal) conventions, see Fantham (1975).

[16] Aristophanes: *Ekklesiazusai*: 383–7.

[17] Xenophon, *Oikonomikos*: 10.2; Aristophanes, *Thesmophoriazusai* 821ff.

7 PERSONAL RELATIONSHIPS

[1] See Lacey (1968: 172–4) for further examples.

[2] For a full description of fourth-century funerary art, see Robertson (1975: 363–82) on whose work the following account is based.

Women are also well represented in funerary inscriptions. Humphreys (1983: 111ff. and 128ff.) has analysed around 600 fourth-century inscriptions with the following results.

234 to individual men
102 to individual women
88 to husband and wife together
11 (plus 2 doubtful) to husband and wife with son(s) and daughter(s)
24 (plus 5 doubtful) to husband and wife with son(s) alone
15 (plus 8 doubtful) to husband and wife with daughter(s) alone
30 (plus 13 doubtful) to father and son(s)
2 (plus 1 doubtful) to mother and son
4 (plus 2 doubtful) to father and daughter
5 (plus 4 doubtful) to mother and daughter
7 (plus 3 doubtful) to brother and brother
5 (plus 1 doubtful) to brother and sister.

A further thirty-seven stones record larger groups of relatives, for the analysis of which see Humphreys (1983). On the above evidence, however, it is clear that while more men are recorded than women, the setting-up of inscriptions to women and, indeed, to women on their own, could be deemed a normal practice.

[3] For a description of the monument to Hagnostrate, daughter of Theodotos, see Robertson (1975: 381). The indications are that Hagnostrate died unwed. For Hegeso, daughter of Proxenos, see Robertson (1975: 367, 370), and for the last, Robertson (1975: 380–1).

[4] The relief also includes a girl looking down at the seated man, and behind him a veiled woman holding the hand of a little girl. See Robertson (1975: 380).

[5] For a description of the monument, see Robertson (1975: 380).

[6] See Dover (1978: 33); the speaker does, however, show embarrassment at having to admit his infatuation for the boy, which would be thought foolish in a man of his age.

[7] But cf. Dover's longer discussion of this question (1972: 160–1): it may be that in ignoring the existence of prostitutes (and of homosexual relations) Aristophanes is suppressing elements that would spoil his plot.

[8] See Dover (1978: 1ff.) for some remarks on this elusive problem.

[9] In treating Athenian (and Greek) homosexuality, I am forced to deal summarily with what is a complex phenomenon. For a full account of the subject, the reader is advised to turn to Dover's excellent work, *Greek Homosexuality* (1978). For the interchangeability of male and female erotic partners, see Dover (1978: 60–8).

[10] In this particular case, no such monument was ever erected. Ischomachos predeceased his wife who subsequently became the (notorious) wife of Kallias, mentioned in Andokides' *On the Mysteries*. See Harvey (1984).

8 THE ATTRIBUTES OF GENDER

1 For a representative selection of examples, and commentary on them, see Dover (1974: 95–102) to whose work I am indebted throughout this chapter.

2 For examples, see Dover (1974: 100–1).

3 Lefkowitz (1986: 10), however, has recently pointed out how difficult it is to ascertain the meaning of such a fragment known only out of context.

4 For discussion of the possible connection between Plato's *Republic 5* and the *Ekklesiazusai*, see Usscher (1973: xiii–xx).

5 It should also be borne in mind that as a result of the Peloponnesian War the Acharnians were starving. The substitution of 'stomach' for 'mind' thus relates to their immediate circumstances; nevertheless, the general connection between females and a concern for gustatory (rather than intellectual) matters is still, I think, being appealed to.

6 For discussion (and further examples) of the 'masculinization' of tragic heroines, see again Pomeroy (1975: 98ff.); cf. Dover (1974: 99).

7 See the extensive writings of M.I. Finley on slavery, and especially 'Between slavery and freedom' reprinted in R.P. Saller and B.D. Shaw (1983), *Economy and Society in Ancient Greece*.

8 See G.E.M. de Ste Croix (1981: 141–2) for the connection between increased rights of the citizen body and increased exploitation of the unfree.

9 I borrow this example from de Ste Croix (1981: 181, 184).

10 See Demosthenes 57 [*Euboulides*] where the speaker's mother had been forced to hire herself out as a wet-nurse. This was used as evidence before a *democratic* court that the speaker's mother was not an *aste*, a citizen woman, and therefore that the speaker himself was not of citizen status. Here a legal classification is actually being deduced from its secondary manifestations – the mode of life which it ideally implied and which was not compatible with the subordination of hired labour. The speaker, of course, rejects the notion that his mother's employment had anything to do with her civic status; nevertheless he admits that she had been (like others) forced into 'slavish occupations' (*doulika pragmata*).

11 In Aristotle's view, differences between the *psyche* of free and slave were matched by physical differences:

> It is then part of nature's intention to make the bodies of free men to differ from those of slaves, the latter strong enough for the necessary menial tasks, the former erect and useless for that kind of work, but well suited for the life of a citizen of a state, a life divided between peace and war.
>
> (Aristotle, *Politics*: 1254b; T.A. Sinclair trans.)

Cf. Dover 'what kind of character expressed the view (Alexis fr. 263) that an awkward and undignified gait is a mark of the "un-free", we do not know, but the point presumably is that the free man should suggest even by his physical movements that he is, as it were, in control of the situation' (1974: 115).

[12] Such an opinion was by no means Aristotle's alone. As Andrewes remarks, 'If a Greek were asked what distinguished his own nation from the rest, one likely reply would be that the Greeks were free and barbarians were slaves' (1971: 273). Aristotle himself cites Homer ('the poet') as an authority.

[13] For further examples, see Dover (1974: 208).

[14] Aischines' prosecution of Timarchos and the legal and social issues relevant to its understanding are fully explored by Dover (1978: chap.2, 19–109), whom I follow in the brief account below.

[15] For a fairly recent discussion of the place of the emotions in the writings of both Plato and Aristotle, see Fortenbaugh (1975).

[16] For the implications of Aristotle's psychological theories for women, children, and slaves, see Fortenbaugh (1975: 45–61).

[17] For a fuller discussion of Aristotle's views, again see Fortenbaugh (1975: 45–61).

9 THE ENEMY WITHIN

[1] For discussion of Euripides' reputation in antiquity as a mysogynist and for an assessment of his portrayal of women, see Pomeroy (1975: 103–12), Vellacott (1975: 82–126) and Lefkowitz (1981: 5–9).

[2] It is, of course, interesting to note that Kreousa's confidant and adviser in this scheme is male; but it should also be stressed that he is a slave, not a free man. The conspiracy is between the slavish and the feminine, whose 'natural' affinities have been discussed in Chapter 8.

[3] See Lloyd-Jones (1959) for a summary of the arguments concerning the end of *Seven Against Thebes* and for a strong case for its integrity.

[4] Despite the conflict between 'natural law' and 'political' edicts represented in Sophokles' play as a conflict between man (Kreon) and woman (Antigone), Antigone herself, like many of the tragic heroines, assumes markedly masculine characteristics – characteristics commented on by Kreon and emphasized by the language of the play, which employs masculine gender adjectives to describe her (see Pomeroy 1975: 99–103). What Greek tragedians had to account for, of course, was the very fact of women assuming an active role in affairs and of being individually capable of challenging male authority. But if an overt 'masculinization' of the heroines (cf. Aischylos' Klytaimnestra) could serve to explain this departure from the representation of reality, this does not deny their simultaneous embodiment of cultural assumptions about the nature and potential dangers of femininity. See also Lefkowitz's discussion of Antigone – that in fact her actions as a woman are not unconventional in their defence of established custom (1986: 81–4).

[5] Goheen's detailed study of the language and imagery of the *Antigone* (1951) is relevant here, for Antigone's own language is marked not

only by a direct emotionalism which couches issues in uncompromisingly black and white terms, but also by a degree of illogicality which actually distances her from the other characters none of whom seem fully to understand her. Antigone may be 'identified with nature and its abiding surety' (1951: 81), but this identification also entails a degree of 'unreasonableness' which even those less speciously rational than Kreon could not hope to broach.

6 As Segal writes, the 'complex connection between control and human relations has also a further significance for Antigone. Her womanly "nature", centred on "sharing love", opposes Creon's attitude of domination which stands apart from the sterness both of men and nature and looks upon them as a potential "enemy" to be subjugated. Thus it is Antigone, the woman – or, perhaps, at another level, the "woman" in him – that Creon must subdue, or, in one of his favourite metaphors, must "yoke" ' (1964: 54). Segal's qualification – 'or, perhaps . . . the "woman" in him' – is revealing, for although Sophokles may use a male/female dichotomy in the conflict he describes, we are not left to feel that the play itself asserts or celebrates any such simple dichotomy.

7 For the position of women in Sparta, see Cartledge (1981: 84–105) and Redfield (1977/8:146–61).

8 For a general discussion of women in war, see Schaps (1982: 193–213).

9 For the full details concerning this case, see Wyse (1904: 232ff.).

10 THE SAVAGE WITHOUT

1 For the location (and reconstruction) of the Homeric world, see Finley (1972); cf. Snodgrass (1974).

2 Cf. Ehrenberg (1974b: 6–7) and Finley (1972: 17–29).

3 See also *Odyssey*: 1.32ff; 3.253ff; 11.405ff.

4 Nausikaa's father, Alkinoos, does mention marriage, and though the proposal comes to nothing, Odysseus still competes in an athletic contest with the young men of the land, thus suggesting the conflation of a story in which the princess was the victor's prize.

5 Gould (1980: 53 n.112) cites *The Bad-tempered Man (Dyskolos)*: 842ff., *The Unkindest Cut (Perikeiromene)*: 1013ff., *The Samian Woman (Samia)*: 726ff.

6 E.g. Aischylos, *Agamemnon*: 1063–6; Anakreon no. 417 in Denys Page (ed.), *Poetae Melici Graeci*, (and for a translation, see the Loeb *Lyra Graeca*, vol. 2. (ed. Edmonds): 181; Euripides, *Andromache*: 227f.; Euripides, *Trojan Women*: 924; Euripides, *Helen*: 357; Euripides, *Hippolytos*: 1148; Sophokles, *Oidipos Tyrannos*: 1485, 1497ff., 1257, 1211ff.; Sophokles, *Antigone*: 569; Sophokles, *Women of Trachis*: 31ff. See Gould (1980: 53) for further examples and references.

7 For an account of the festival of Artemis at Brauron, see the writings of Kahil (1965, 1977, and particularly 1983). The age of the participants admits of some speculation, and perhaps it ranged up to fourteen years. See Perlman (1983).

[8] Aelian, *Varia Historia*: 13.1.

[9] The transition of mankind from a blessed pre-agricultural age to one in which men must labour for their sustenance is brilliantly analysed by Vernant in two essays on Hesiod (1981a and 1981b). The connection between this transition and the advent of marriage – indeed of woman herself – again forms part of the complex set of associations worked through by Hesiod's account.

[10] Apollodoros, *Library*: 3.9.2.

[11] Hyginus, *Fabulae*: 185; Servius, commenting on Virgil, *Aeneid*: 3.113. The ancient mythographers refer their explanation of the punishment to Pliny's *Natural History*: (8.43), but in fact Pliny's account of the mating habits of lions does not support their interpretation (see Frazer's note in the Loeb edition of *Apollodorus*, vol. I, (1921: 401 n.2)).
For further comment on the significance of the Atalanta myth, see Vidal-Naquet (1981b: 161–2). Cf. Tyrrell (1984: 83–4).

[12] See above n.11 and, for example, Hyginus, *Fabulae*: 185, Theokritos 2.1. 40–2; Kallimachos, *Hymn to Artemis*: 215f.; Pausanias 8.45; Ovid, *Metamorphoses*: 10.560ff.; Diodoros 4.34. Most of the sources are late or fragmentary, but clearly the myth was well known in fifth-century Athens; Atalanta was mentioned in a lost tragedy by Euripides, while Melanion is mentioned in Aristophanes' *Lysistrata*: 785–95. The names of Atalanta's father and of her captor/husband are, however, anything but constant in the tradition. See Frazer vol.1 (1921: 398).

[13] For the myth of Peleus and Thetis see, for example, Apollodoros, *Library*: 3.13.4ff. and Ovid, *Metamorphoses*: 11.213ff. For further sources, see Frazer, vol.2 (1921: 66ff.).

[14] See again Vernant's essays on Hesiod (1981a: 43ff. and 1981b: 57ff.).

[15] See Tyrrell for the role of Hippolytos as the liminal male who fails to mature. 'Hippolytus is a myth, a figment of the failure to mature. Compared to the father defined by the Amazon myth's structure, he is socially useless: his power to reproduce does not contribute to the family or state. What better image of his condition than Euripides' sexless human wandering the wilds with an immortal virgin who shuns men and is invisible even to a favourite? His eroticism, on the other hand, operating in the extant play as the passion aroused in the married woman Phaedra, is outside marriage and thus open for Aphrodite to use destructively. She – that is, Hippolytus' erotic appeal – destroys Theseus' marriage as well as Phaedra, Theseus, and Hippolytus himself' (1984: 85). Cf. Vidal-Naquet (1981b: 160).

[16] Pausanias 1.28.5; cf. Ovid, *Heroides*: 4.67ff. Even in Euripides' play Phaidra dreams of joining Hippolytos in the wilds and the hunt (108ff.).

[17] For Pasiphai and the bull, see Apollodoros, *Library*: 3.1.4 and 3.15.8, and Ovid, *The Art of Love*: 1.289ff. For Leda and the Swan, see Apollodoros, *Library*: 3.10.6–7. For Europa and the bull, see Apollodoros, *Library*: 3.1.1, and Ovid, *Metamorphoses*: 2.836ff. and *Fasti*: 5. 603ff. The sources cited above, which give a relatively sustained account of the stories, are, of course, very late. But again the myths were well known in fifth- century Athens. Euripides' lost play, *The*

Cretans, appears to have dealt with Pasiphai's passion; Helen briefly recounts the story of Leda (her mother) and the swan in Euripides' play, *Helen*. For the many fragmentary mentions of Pasiphai, Leda and Europa, see the notes in the Loeb edition of *Apollodorus* (Frazer 1921), or the references in Grimal's *Dictionary of Classical Mythology* (1986).

18 For the locations of the Amazons in specific sources, see Tyrrell (1984: 142 n.45). Homer, for example, places them in Lycia and Phrygia (*Iliad*: 6.172–86); Diodoros has them as far removed as Libya (3.53.1). The majority of sources locate them on the southern shore of the Black Sea.

19 Strabo 11.5.1; Philostratos, *Heroikos*: 330 (Teubner edn, vol. 2: 216); Diodoros 3.53.2. Herodotos (4.110–17) maintains their initial promiscuity, but they settle down to permanent marriage with the Skythians.

20 Strabo 11.5.1; Diodoros 2.45.3; Philostratos, *Heroikos*: 330 (Teubner edn, vol. 2: 216); Hippokrates, *On Joints*: 53.

21 Apollodoros, *Library* 2.5.9; Diodoros 2.45.3; Strabo 11.5.1. Arrian (*Anabasis*: 7.13.2) says only that their right breast was smaller, and in any case doubts the authenticity as Amazons of the women whom he is describing. See Tyrrell (1984: 140 n. 18) and Engle (1942: 514–15) for further references.

22 Philostratos, *Heroikos*: 330 (Teubner edn, vol. 2: 216–17).

23 Strabo 11.5.1; Diodoros 3.54.2; Philostratos, *Heroikos*: 330 (Teubner edn, vol. 2: 216); cf. Euripides, *Madness of Herakles*: 408f.; Pindar *Olympian Odes*: 8.47; Aristophanes, *Lysistrata*: 676–7. See Engle (1941: 516), some of whose references, however, are inaccurate.

24 Apollonios of Rhodes, *Argonautika*: 2. 987–9.

25 Diodoros 3.53.2.

26 The theory of primitive matriarchies was first expounded (in academic form) by Bachofen in *Das Mutterecht (Mother Right)* in 1861. A somewhat similar thesis concerning the existence of matriarchies within the evolution of society was also presented by Morgan in *Ancient Society* (1877). Since later scholarship, both historical and ethnographic, has singularly failed to substantiate the existence of matriarchies, the whole issue might well have died were it not for its romantic (and mystical) support by popular writers from Robert Graves to Mary Renault, and the fact that both Bachofen and Morgan were canonized by their inclusion in Engels' *The Origin of the Family, Private Property, and the State* (1884), thus ensuring the survival of matriarchies in some Marxist-feminist writings.

27 For the 'effeminacy' of barbarians (and its connections with notions both of slavery and savagery, as well as geography) see Tyrrell's excellent discussion (1984: 60–3).

28 Diodoros 3.53. 5ff., and see Tyrrell (1984: 57–9).

29 See Tyrrell (1984: 54, 66, 78–81).

30 Strabo 11.5.1; cf. Philostratos, *Heroikos:* 330.

31 This is their standard parentage. See, for example, Isokrates 4.68 and Lysias 2.4.

32 Strabo 11.5.4; Pausanias 4.31.8. However, Pausanias contradicts himself in 7.2.7 where he specifically denies that the Amazons founded the sanctuary.

33 Pausanias 3.25.3.

34 Apollodoros, *Epitome* 1.21.

35 Plutarch (*Life of Theseus*: 26), who quotes earlier sources, claims that Theseus made an expedition against the Amazons separate from Herakles' (see below) and with his own companions. Plutarch does not mention Peirithous, but in Pausanias 1.21. it is related (on the authority of Pindar) that Theseus and Peirithous together carried off the Amazon queen. It could be assumed, then, that Peirithous was Theseus' companion on the expedition.

36 For Centaurs, see Kirk (1970: 152–62).

37 In Aischylos' *Suppliant Women* (287) they are described as 'manless flesh-eating Amazons' – though the adjective does not make it certain that they devoured their flesh raw, and may simply refer to their ignorance of agriculture (cf. Diodoros 3.53.5). Even that, however, likens them to the Centaurs in their lack of civilization.

38 Strabo 11.51. This is not, however, the clothing in which they are usually depicted. See n. 48 below.

39 Euripides, *Madness of Herakles*: 418.

40 Apollonios of Rhodes, *Argonautika*: 2. 966–9; Diodoros 4.16.

41 Plutarch, *Life of Theseus* 26.1; Diodoros 4.16.

42 Diodoros 3.55.3.

43 See, for example, Thoukydides 2.15; Isokrates 10 [*Helen*]: 32–8; Euripides, *Suppliants*: 403–8.

44 For a full account of the promotion and interpretation of Theseus in classical Athens, see Connor (1970).

45 Plutarch, *Life of Theseus*: 26.1–2 and 28.1; Pausanias 1.2.1; Isokrates 12 [*Panathenaikos*]: 193; Diodoros 4.28.3.

46 Plutarch, *Life of Theseus*: 27; Isokrates 12 [*Panathenaikos*]: 193, and 4 [*Panegyrikos*]: 68–70; Lysias 2 [*Funeral Oration*]: 4–6.

47 Plutarch, *Life of Theseus*: 27.4; Diodoros 4.28.4.

48 See Thoukydides 9.27.1–5. See Tyrrell (1984: 13–19) for a good discussion of the use of Amazons in funeral oratory. In the classical period, the Amazons are standardly assimilated with the Persians by their dress and weaponry (Tyrrell 1984: 49–52).

49 In what follows, I am greatly indebted to the interpretations of Segal (1984) for the *Bacchai*, and Shaw (1975) for the *Medea*.

50 See Segal's excellent comments (1984: 199–200) on Euripides' stance and the role of tragedy as standing critically outside the norms of Athenian culture.

51 Cf. Segal: 'The "problem of the Bacchae", then, has no resolution; and the power of the tragedy lies in the vehemence with which the two sides clash and in the untimigated horror of the wreckage that emerges from the encounter' (1984: 209).

[52] The following is based largely on Apollodoros, *Library* 1.9.23–8 and *Epitome*: 1.5–7, in itself a late and synthetic account. Other major sources for Medea's career include Apollonios of Rhodes, *Argonautika* (for the most part followed by *Apollodoros*); Diodoros 4.46.1–4.56.2; Plutarch, *Life of Theseus*; and, of course, Euripides' *Medea*.

[53] Apollodoros' *Library*: 1.9.1 for Medea's kinship with Kirke and Pasiphai, and Apollonios of Rhodes, *Argonautika*: 4.682–4 for Kirke and Aietes as brother and sister.

[54] Apollodoros, *Library*: 1.9.23; Apollonios, *Argonautika* 3.401ff. and 3.1176ff. The dragon's teeth sprout armed men; Jason deals with them in the same way as Kadmos did at Thebes – by throwing stones among them so that they set upon each other.

[55] Thus Apollodoros (*Library*: 1.9.23–4), but this is not the version given by Apollonios, *Argonautika*: 4.224–481, in which Apsyrtos leads the Kolchian fleet in pursuit of the Argonauts and is then treacherously killed by Jason (with Medea's help) when they establish a truce. In a lost play by Sophokles (*The Kolchian Women*) it seems that Apsyrtos was killed by Medea in the palace of Aietes, and this appears the version presented in Euripides, *Medea*: 1334. For further sources see Frazer, vol.1, (1921:113).

[56] For Jason's claim to the throne of Iolkos, and its usurpation by Pelias, see Pindar, *Pythian Odes* 4.106ff. For Aison's death, see Apollodoros' *Library*: 1.9.27 and Diodoros 4.50.1 in which he is forced (or chooses) to drink bull's blood.

[57] Apollodoros, *Library*: 1.9.27–8; Diodoros 4.53.51–4. Diodoros' version is more complex. Medea employs a range of magical tricks, and Jason goes voluntarily to Corinth.

[58] Apollodoros' *Library* follows the version given by Euripides, but notes the version in which they were killed by the Corinthians. See Pausanias 2.3.6 for the annual expiations.

[59] Apollodoros, *Library*: 1.9.28 and *Epitome*: 1.5–6; Diodoros 4.55.4–6; Plutarch, *Life of Theseus*: 12; Pausanias 2.3.8.

[60] Apollodoros, *Library*: 1.9.28; Diodoros 4.56.1–2. Medea's eventual union with Achilles relies on the scholiast on Euripides, *Medea*: 10 and on Apollonios, *Argonautika*: 4.814.

[61] The accounts of Medea's career in Apollodoros, Apollonios, and Diodoros are replete with further examples of her magic and use of drugs and potions.

[62] Apollodoros, *Library*: 3.15.1.

[63] Rather lurid accounts of Kirke's magic are also given in Apollonios, *Argonautika*: 4.662ff. and Diodoros 4.45. 3–4.

[64] Sophokles, *Women of Trachis*: 1136ff.

[65] Pindar, *Pythian Odes*: 4.214ff.

[66] Theokritos 2.

[67] On the authority of the scholiast on *Medea*: 9, it seems accepted that Medea's killing of her own children was Euripides' invention.

[68] Cf. Shaw (1975: 263–4). Shaw, however, argues (1975: 262ff.) that from the point at which Medea decides to avenge herself on Jason she

becomes masculine and that the conflict between her *thymos* and her *bouleumata* ('plans', 'rational thought') is one between her 'masculine' and 'feminine' self. Medea, like most heroines of Greek tragedy, does take on markedly 'masculine' characteristics if only through the forcefulness of her actions. I would not, however, go so far as to construe *thymos* as specifically '*male* heart'. Nor, in terms of Athenian ideology, can I see much cause for associating 'rational thought' with femininity. The conflict between rationality and passion is certainly there, but the triumph of passion over rational thought, of *thymos* over *bouleumata*, is consistent with Athenian views about the female *psyche*, even if, outside of tragedy, women had little opportunity to demonstrate the power of their *thymos*.

BIBLIOGRAPHY

TRANSLATIONS

The translations of Greek sources quoted in this book have been based on the following works. Some (for example Lattimore's translation of the *Odyssey*) have scarcely been changed. Others (for example the older Loeb translations of the Orators) have been substantially modified. In some very few cases I have offered my own translations.

Aischines, 1 [*Timarchos*], from *The Speeches of Aeschines*, trans. C.D. Adams, Loeb Classical Library, London: William Heinemann, 1919.

Aischylos, *Agamemnon*, trans. R. Lattimore, from *Aeschylus I, Oresteia*, in D. Grene and R. Lattimore (eds), *The Complete Greek Tragedies*, University of Chicago Press, 1953.

Aischylos, *Eumenides*, trans. R. Lattimore, from *Aeschylus I, Oresteia*, in D. Grene and R. Lattimore (eds), *The Complete Greek Tragedies*, University of Chicago Press, 1953.

Aischylos, *Seven Against Thebes*, H. Weir Smyth, Loeb Classical Library, London: William Heinemann, 1922.

Antiphon, 1 [*Stepmother*], from *Minor Attic Orators I*, trans. K. J. Maidment, Loeb Classical Library, London: William Heinemann, 1941.

Apollodoros, *Library*, from *Apollodorus* (2 vols), trans Sir J. G. Frazer, Loeb Classical Library, London: William Heinemann, 1921.

Aristophanes, *Lysistrata*, from *Aristophanes III*, trans. B.B. Rogers, Loeb Classical Library, London: William Heinemann, 1924, and from K.J. Dover, *Aristophanic Comedy*, Berkeley and Los Angeles: University of California Press, 1972.

Aristophanes, *Wasps* from *Aristophanes I*, trans. B.B. Rogers, Loeb Classical Library, London: William Heinemann, 1924.

Aristotle, Al-Dailami, cod. Tubingen Weisweiler, from *The Works of Aristotle Translated into English, Vol XII Select Fragments*, ed. Sir D. Ross, Oxford: Clarendon Press, 1952.

Aristotle, *Nikomachean Ethics*, from *The Ethics of Aristotle*, trans. J.A.K. Thomson, Harmondsworth: Penguin, 1953
Aristotle, *Politics*, from *Aristotle, The Politics*, trans. T.A. Sinclair, Harmondsworth: Penguin, 1962.
Demosthenes, 18 [*On the Crown*], from *Demosthenes II*, trans. C.A. Vince and J.H. Vince, Loeb Classical Library, London: William Heinemann, 1926.
Demosthenes, 19 [*Embassy*], from *Demosthenes II*, trans. C. A. Vince and J. H. Vince, Loeb Classical Library, London: William Heinemann, 1926.
Demosthenes, 23 [*Aristokrates*], from *Demosthenes III*, trans. J.H. Vince, Loeb Classical Library, London: William Heinemann, 1935.
Demosthenes, 25 [*Aristogeiton*], from *Demosthenes III*, trans J.H. Vince, Loeb Classical Library, London: William Heinemann, 1935.
Demosthenes, 30 [*Onetor*], from *Demosthenes IV*, trans. A. T. Murray, Loeb Classical Library, London: William Heinemann, 1936.
Demosthenes, 39 [*Boiotos I*] from *Demosthenes IV*, trans. A.T. Murray, Loeb Classical Library, London: William Heinemann, 1936.
Demosthenes, 40 [*Boiotos II*], from *Demosthenes IV*, trans. A.T. Murray, Loeb Classical Library, London: William Heinemann, 1936.
Demosthenes, 43 [*Makartatos*], from *Demosthenes V*, trans. A.T. Murray, Loeb Classical Library, London: William Heinemann, 1939.
Demosthenes, 46 [*Stephanos II*], from *Demosthenes V*, trans. A.T. Murray, Loeb Classical Library, London: William Heinemann, 1939.
Demosthenes, 48 [*Olympiodoros*], from *Demosthenes V*, trans. A.T. Murray, Loeb Classical Library, London: William Heinemann, 1939.
Demosthenes, 50 [*Polykles*], from *Demosthenes VI*, trans. A.T. Murray, Loeb Classical Library, London: William Heinemann, 1939.
Demosthenes, 54 [*Konon*], from *Demosthenes VI*, trans. A.T. Murray, Loeb Classical Library, London: William Heinemann, 1939.
Demosthenes, 55 [*Kallikles*], from *Demosthenes VI*, trans. A.T. Murray, Loeb Classical Library, London: William Heinemann, 1939.
Demosthenes, 57 [*Euboulides*], from *Demosthenes VI*, trans. A.T. Murray, Loeb Classical Library, London: William Heinemann, 1939.
Demosthenes, 59 [*Neaira*], from *Demosthenes VI*, trans. A.T. Murray, Loeb Classical Library, London: William Heinemann, 1939.
Demosthenes, 60 [*Funeral Oration*], from *Demosthenes VII*, trans. N.W. De Witt and N.J. De Witt, Loeb Classical Library, London: William Heinemann, 1949.
Demosthenes, 61 [*Erotic Essay*], from *Demosthenes VII*, trans. N.W. De Witt and N.J. De Witt, Loeb Classical Library, London: William Heinemann, 1949.
Euripides, *Bacchai*, trans. W. Arrowsmith, from *Euripides V*, in D. Grene and R. Lattimore (eds), *The Complete Greek Tragedies*, University of Chicago Press, 1959.
Euripides, *Hippolytos*, extract trans. in K.J. Dover, *Greek Popular Morality*, Oxford: Basil Blackwell, 1974, and trans. D. Grene, from *Euripides I*, in

D. Grene and R. Lattimore (eds), *The Complete Greek Tragedies*, University of Chicago Press, 1955.
Euripides, *Medea*, trans. R. Warner, from *Euripides I*, in D. Grene and R. Lattimore (eds), *The Complete Greek Tragedies*, University of Chicago Press, 1955.
Homer, *Odyssey*, trans. R. Lattimore, *The Odyssey of Homer*, New York: Harper and Row, 1965.
Hypereides, 1 [*Lykophron*), from *Minor Attic Orators II*, trans. J.O. Burtt, Loeb Classical Library, London William Heinemann, 1954.
Isaios, 2 [*Menekles*], from *Isseus*, trans. E.S. Forster, Loeb Classical Library, London: William Heinemann, 1927.
Isaios, 3 [*Pyrrhos*], from *Isaeus*, trans. E.S. Forster, Loeb Classical Library, London: William Heinemann, 1927.
Isaios, 6 [*Philoktemon*], from *Isaeus*, trans. E.S. Forster, Loeb Classical Library, London: William Heinemann, 1927.
Isaios 7 [*Apollodoros*], from *Isaeus*, trans. E.S. Forster, Loeb Classical Library, London: William Heinemann, 1927.
Isaios, 8 [*Kiron*], from *Isaeus*, trans. E.S. Forster, Loeb Classical Library, London: William Heinemann, 1927.
Isaios, 9 [*Astyphilos*], from *Isaeus*, trans. E.S. Forster, Loeb Classical Library, London: William Heinemann, 1927.
Isaios, 10 [*Aristarchos*], from *Isaeus*, trans. E.S. Forster, Loeb Classical Library, London: William Heinemann, 1927.
Isaios, 11 [*Hagnias*], from *Isaeus*, trans. E.S. Forster, Loeb Classical Library, London: William Heinemann, 1927.
Isaios, 12 [*Euphiletos*], from *Isaeus*, trans. E.S. Forster, Loeb Classical Library, London: William Heinemann, 1927.
Isokrates, 14 [*Panegyrikos*] from *Isocrates I*, trans. G. Norlin, Loeb Classical Library, London: William Heinemann, 1928.
Isokrates, 12 [*Panathenaikos*] from *Isocrates II*, trans. G. Norlin, Loeb Classical Library, London: William Heinemann, 1928.
Lysias, 1 [*Eratosthenes*], from *Lysias*, trans. W.R.M. Lamb, Loeb Classical Library, London: William Heinemann, 1930.
Lysias, 2 [*Funeral Oration*], from *Lysias*, trans. W.R.M. Lamb, Loeb Classical Library, London: William Heinemann, 1930.
Lysias, 3 [*Simon*], from *Lysias*, trans. W.R.M. Lamb, Loeb Classical Library, London: William Heinemann, 1930.
Lysias, 4 [*On a Wounding*], from *Lysias*, trans. W.R.M. Lamb, Loeb Classical Library, London: William Heinemann, 1930.
Lysias, 21 [*Defence against a Charge of Taking Bribes*], from *Lysias*, trans. W.R.M. Lamb, Loeb Classical Library, London: William Heinemann, 1930.
Lysias, 32 [*Diogeiton*], from *Lysias*, trans. W.R.M. Lamb, Loeb Classical Library, London: William Heinemann, 1930.
Menander, *Dyskolos (Bad-Tempered Man)*, from *Theophrastus, The Characters, Menander Plays & Fragments*, trans, P. Vellacott, Harmondsworth: Penguin, 1973.

Menander, *Perikeiromene* (*The Unkindest Cut*), from *Theophrastus, The Characters, Menander Plays & Fragments*, trans. P. Vellacott, Harmondsworth: Penguin, 1973.
Plato, *Laws*, from *Plato, Laws*, trans. T.J. Saunders, Harmondsworth: Penguin, 1970.
Plato, *Timaios*, from *The Dialogues of Plato Vol. 3*, trans. B. Jowett, London: Sphere Books, 1970.
Plutarch, *Perikles*, from *Plutarch, The Rise and Fall of Athens*, trans. I. Scott-Kilvert, Harmondsworth: Penguin, 1960.
Sophokles, *Antigone*, trans. E. Wyckoff, from *Sophocles I*, in D. Grene and R. Lattimore (eds), *The Complete Greek Tragedies*, University of Chicago Press, 1954.
Sophokles, *Trachiniai (The Women of Trachis)*, trans M. Jameson, from *Sophocles II*, in D. Grene and R. Lattimore (eds), *The Complete Greek Tragedies*, University of Chicago Press, 1957.
Thoukydides, *Thucydides, History of the Peloponnesian War*, trans. Rex Warner, Harmondsworth: Penguin, 1954.
Xenophon, *Memorabilia*, from *Xenophon, Memorabilia and Oeconomicus*, trans. E.C. Marchant, Loeb Classical Library, London: William Heinemann, 1923.
Xenophon, *Oikonomikos*, from *Xenophon, Memorabilia and Oeconomicus*, trans. E.C. Marchant, Loeb Classical Library, London: William Heinemann, 1923.
Xenophon, *Symposion*, from *Xenophon Anabasis IV–VII, Symposium, Apology*, trans, C.L. Brownson and O.J. Todd, Loeb Classical Library, London: William Heinemann, 1922.

GENERAL BIBLIOGRAPHY

Alexiou, M. (1974) *The Ritual Lament in Greek Tradition*, Cambridge: Cambridge University Press.
Andrewes, A. (1971) *Greek Society*, Harmondsworth: Penguin.
Ardener, E. (1975) 'Belief and the problem of women', in S. Ardener (ed.) (1975). (First published in J. S. La Fontaine (ed.) (1972) *The Interpretation of Ritual*, London: Tavistock).
Ardener, S. (ed.) (1975) *Perceiving Women*, London: Malaby Press.
Ardener, S. (ed.) (1978) *Defining Females*, London: John Wiley & Sons.
Ardener, S. (ed.) (1981) *Women and Space*, London: Croom Helm.
Arthur, M. B. (1984) 'The origins of the western attitude towards women', in J. Peradotto and J. P. Sullivan (eds) (1984). (First published in *Arethusa*, 6: 1–58.)
Benveniste, E. (1969) *Le Vocabulaire des Institutions Indo-Europeénnes, Vol. I. Economie, Parente, Société*, Paris: Les Editions de Minuit.

Bickerman, E. J. (1975) 'La conception du marriage à Athènes', *Bullettino Dell'Istituto di Diritto Romano*, 78: 1–28.
Bonner, R. G. and Smith, G. (1938) *The Administration of Justice from Homer to Aristotle, Vol. II,* Chicago: University of Chicago Press.
Cameron, A. and Kuhrt, A. (eds) (1983) *Images of Women in Antiquity*, London: Croom Helm.
Carteledge, P. (1981) 'Spartan wives: liberation or licence', *Classical Quarterly* 31: 84–105.
Connor, W. R. (1970) 'Theseus in classical Athens', in A. G. Ward (ed.) *The Quest for Theseus*, New York: Praeger.
Davies, J. K. (1971) *Athenian Propertied Families 600–300 BC,* Oxford: Clarendon Press.
de Ste Croix, G. E. M. (1981) *The Class Struggle in the Ancient World*, London: Duckworth.
Detienne, M. (1977) *The Gardens of Adonis. Spices in Greek Mythology*, Sussex: The Harvester Press. (First published in France as *Les Jardins d'Adonis*, Paris: Editions Gallimard, 1971.)
Dodds, E. R. (1951) *The Greeks and the Irrational*, Berkeley and Los Angeles: University of California Press.
Dover, K. J. (1972) *Aristophanic Comedy*, Berkeley and Los Angeles: University of California Press.
Dover, K. J. (1974) *Greek Popular Morality in the Time of Plato and Aristotle*, Oxford: Basil Blackwell.
Dover, K. J. (1978) *Greek Homosexuality*, London: Duckworth.
Dover, K. J. (1984) 'Classical Greek attitudes to sexual behaviour', in J. Peradotto and J. P. Sullivan (eds) (1984). (First published in *Arethusa* 6: 59–73).
du Boulay, J. (1986) 'Women – images of their nature and destiny in rural Greece', in J. Dubisch (ed.) (1986).
Dubisch, J. (ed.) (1986) *Gender and Power in Rural Greece*, Princeton: Princeton University Press.
Ehrenberg, V. (1974a) *The People of Aristophanes*, London: Methuen. (First published in 1943, Oxford: Basil Blackwell.)
Ehrenberg, V. (1974b) *Man, State and Deity*, London: Methuen.
Engle, B. S. (1942) 'The Amazons in ancient Greece', *The Psychoanalytic Quarterly* 11: 512–54.
Fantham, E. (1975) 'Sex, status and survival in Hellenistic Athens', *Phoenix* 29: 44–74.
Ferguson, W. S. (1910) 'The Athenian phratries', *Classical Philology* 5: 257–84.
Fine, A. J. V. (1951) 'Horoi: studies in mortgage, real security, and land tenure in ancient Athens', *Hesperia* Supplement IX.
Finley, M. I. (1951) *Land and Credit in Arcient Athens 500–200 BC*, New Brunswick: Rutgers University Press.
Finley, M. I. (1954) 'Marriage, sale and gift in the Homeric world', *Seminar* 12.
Finley, M. I. (1972) *The World of Odysseus*, Harmondsworth: Penguin.
Finley, M. I. (1973) *The Ancient Economy*, London: Chatto & Windus.

Finley, M. I. (1975) *The Ancient Greeks*, Harmondsworth: Penguin.
Finley, M. I. (1983) 'Between freedom and slavery', reprinted in R. P. Saller and B. D. Shaw (eds) *Economy and Society in Ancient Greece*, London: Chatto & Windus.
Flacelière, E. R. (1965) *Daily Life in Greece at the Time of Pericles*, London: Weidenfeld & Nicholson.
Fortenbaugh, W. W. (1975) *Aristotle on Emotion*, London: Duckworth.
Frazer, J. G. (1921) *Apollodorus. The Library (2 vols), with an English translation by Sir James George Frazer*, Loeb Classical Library, London: William Heinemann.
Friedl, E. (1962) *Vasilika. A Village in Modern Greece*, New York: Holt, Rinehart & Winston.
Fustel de Coulanges Numa Denis (1864) *La Cité Antique*, Paris: Durand. (Now available as *The Ancient City*, Baltimore and London: The Johns Hopkins University Press, 1980.)
Gernet, L. (1920) 'La Création du testament', *Revues des Etudes Grècque*, 33: 123–68 and 249–90.
Gernet, L. (1921) 'Sur l'Epiclerat', *Révue des Etudes Grècques* 34: 337–79
Gernet, L. (1955) *Droit et Société dans la Grèce Ancienne*, Paris: Publications de l'Institut de Droit Romain de l'Université de Paris.
Goheen, R. F. (1951) *The Imagery of Sophocles' Antigone*, Princeton, New Jersey: Princeton University Press.
Gomme, A. W. (1925) 'The position of women in Athens in the fifth and fourth centuries', *Classical Philology* 20: 1–25.
Gordon, R. L. (ed.) (1981) *Myth, Religion and Society*, Cambridge/Paris: Cambridge University Press/Editions de la Maison des Sciences de l'Homme.
Gould, J. (1980) 'Law, custom and myth: aspects of the social position of women in classical Athens, *Journal of Hellenic Studies* 100: 38–59.
Grimal, P. (1986) *The Dictionary of Classical Mythology*, Oxford: Basil Blackwell.
Haigh, A. E. (1907) *The Attic Theatre* (3rd edn revised and in part rewritten by A. W. Pickard-Cambridge), Oxford: Clarendon Press.
Hansen, M. H. (1985) *Demography and Democracy. The Number of Athenian Citizens in the Fourth Century BC*, Denmark: Systime.
Hansen, M. H. (1988) *Three Studies in Athenian Demography. Historisk-filosofiske Meddelelser 56*, Copenhagen: The Royal Danish Academy of Sciences and Letters.
Harrison, A. R. W. (1947) 'A problem in the rules of intestate succession at Athens', *Classical Review* 61: 41–3.
Harrison, A. R. W. (1968) *The Law of Athens Vol. I, The Family and Property*, Oxford: Clarendon Press.
Harrison, A. R. W. (1971) *The Law of Athens Vol. II, Procedure*, Oxford: Clarendon Press.
Harvey, D. (1984) 'The wicked wife of Ischomachos', *Echos du Monde Classique: Classical Views* (n.s.) 3: 68–70.
Harvey, D. (1985) 'Women in Thucydides', *Arethusa* 18: 67–90.
Herfst, P. (1922) *La Travail de la Femme dans la Grèce Ancienne*, Utrecht:

A. Oosthoek.
Hirschon, R. (1978) 'Open body/closed space', in S. Ardener (ed.) (1978).
Hopper, R. F. (1957) 'The Basis of the Athenian Democracy. Inaugural Lecture', Sheffield: Sheffield University Press.
Humphreys, S. C. (1974) 'The nothoi of Kynosarges', *Journal of Hellenic Studies* 94: 88–95.
Humphreys, S. C. (1978) *Anthropology and the Greeks*, London: Routledge & Kegan Paul.
Humphreys, S. C. (1983) *The Family, Women and Death*, London: Routledge & Kegan Paul.
Humphreys, S. C. and Momigliano, A. (1978) 'The social structure of the city', in S. C. Humphreys (1978).
Jaeger, W. (1939) *Paideia: The Ideals of Greek Culture, Vol. I,* Oxford: Basil Blackwell.
Jaeger, W. (1945) *Paideia: The Ideals of Greek Culture, Vol. III,* Oxford: Basil Blackwell.
Jones, A. H. M. (1957) *Athenian Democracy*, Oxford: Basil Blackwell.
Kahil, L. (1965) 'Autour de l'Artémis attique', *Antike Kunst* 8: 20–33.
Kahil, L. (1977) 'L'Artémis de Brauron: rites et mystères', *Antike Kunst* 20: 86–101.
Kahil, L. (1983) 'Mythological repetoire at Brauron', in W. G. Moon (ed.) *Ancient Greek Art and Iconography*, Madison, Wisconsin: University of Wisconsin Press.
King, H. (1983) 'Bound to bleed: Artemis and Greek women', in A. Cameron and A. Kuhrt (eds) (1983).
Kirk, G. S. (1970) *Myth, its Meaning and Function in Ancient and Other Cultures*, Berkeley and Cambridge: University of California Press/ Cambridge University Press.
Kirk, G. S. (1974) *The Nature of Greek Myths*, Harmondsworth: Penguin.
Kitto, H. D. F. (1951) *The Greeks*, Harmondsworth: Penguin.
Kuenen-Janssens, L. J. (1941) 'Some notes on the competence of the Athenian woman to conduct a transaction', *Mnemosyne* (3rd ser.) 9: 199–214.
Lacey, W. K. (1968) *The Family in Classical Greece*, London and New York: Thames & Hudson.
Leach, E. (1961) *Rethinking Anthropology*, London: Athlone Press.
Lefkowitz, M. R. (1981) *Heroines and Hysterics*, London: Duckworth.
Lefkowitz, M. R. (1986) *Women in Greek Myth*, London: Duckworth.
Lévi-Strauss, C. (1969) *The Elementary Structures of Kinship*, London: Eyre & Spottiswoode.
Lloyd-Jones, P. H. J. (1959) 'The end of the *Seven Against Thebes*', *Classical Quarterly* (n.s.) 9: 80–115.
Lowes Dickinson, G. (1938) *The Greek View of Life*, London: Methuen. (First published 1896.)
MacDowell, D. M. (1963) *Athenian Homicide Law in the Age of the Orators*, Manchester: Manchester University Press.
MacDowell, D. M. (1976) 'Bastards as Athenian citizens', *Classical Quarterly* 70: 88–91.

Needham, R. (1971) 'Remarks on the analysis of kinship and marriage', in R. Needham (ed.) *Rethinking Kinship and Marriage*, London: Tavistock.
Nilsson, M. P. (1964) *A History of Greek Religion*, New York: W. W. Norton. (First published in English, Oxford University Press, 1925.)
Padel, R. (forthcoming) *In and Out of the Mind: Consciousness in Greek Tragedy*,
Peçirka, J. (1963) 'Land tenure and the development of the Athenian polis', in L. Varçl and R. F. Willetts (eds) *Geras, Studies Presented to George Thomson on the Occasion of his 60th Birthday,* Prague: Charles University Press.
Peradotto, J. and Sullivan, J. P. (eds) (1984) *Women in the Ancient World. The Arethusa Papers*, Albany: State University of New York Press.
Perlman, P. (1983) 'Plato, *Laws*: 833c–834d and the bears of Brauron', *Greek, Roman and Byzantine Studies* 20: 115–30.
Pomeroy, S. B. (1973) 'Selected bibliography on women in antiquity', *Arethusa* 6: 125–52.
Pomeroy, S. B. (1975) *Goddesses, Whores, Wives and Slaves*, New York: Schocken Press.
Pomeroy, S. B. (1983) 'Infanticide in Hellenistic Greece', in A. Cameron and A. Kuhrt (eds) (1983).
Pomeroy, S. B. (1984) 'Selected bibliography on women in classical antiquity', in J. Peradotto and J. P. Sullivan (eds) (1984).
Post, L. A. (1940) 'Woman's place in Menander's Athens', *Transactions and Proceedings of the American Philological Association* 71: 420–59.
Redfield, J. (1977/8) 'The women of Sparta', *Classical Journal* 73: 146–61.
Rhodes, P. J. (1978) 'Bastards as Athenian citizens', *Classical Quarterly* 72: 89–92.
Richter, C. D. (1971) 'The position of women in classical Athens', *The Classical Journal* 67: 1–8.
Robertson, M. (1975) *A History of Greek Art* (two vols), Cambridge: Cambridge University Press.
Schaps, D. M. (1977) 'The woman least mentioned: etiquette and women's names', *Classical Quarterly* 27: 323–31.
Schaps, D. M. (1979) *Economic Rights of Women in Ancient Greece*, Edinburgh: Edinburgh University Press.
Schaps, D. M. (1982) 'The women of Greece in wartime', *Classical Philology* 77: 193–213.
Segal, C. (1984) 'The menace of Dionysus: sex roles and reversals in Euripides' *Bacchae*', in J. Peradotto and J. P. Sullivan (eds) (1984).
Seltman, C. (1956) *Women in Antiquity*, London and New York: Thames & Hudson.
Shaw, M. (1975) 'The female intruder: women in fifth-century drama', *Classical Philology* 70: 255–66.
Smith, C. (1892) 'Harpies in Greek art', *Journal of Hellenic Studies* 13: 109–14.
Snodgrass, A. M. (1974) 'An historical Homeric society?' *Journal of Hellenic Studies* 94: 114–25.
Sperber, D. (1975) *Rethinking Symbolism*, Cambridge: Cambridge

University Press.
Thompson, W. E. (1967) 'The marriage of first cousins in Athenian society', *Phoenix* 21: 273–82.
Thompson, W. E. (1971) 'Attic kinship terminology', *Journal of Hellenic Studies* 91: 110–13.
Thompson, W. E. (1972) 'Athenian marriage patterns: remarriage', *California Studies in Classical Antiquity* 5: 211–25.
Tucker, T. G. (1907) *Life in Ancient Athens*, London: Macmillan.
Tyrrell, W. B. (1984) *Amazons. A Study in Athenian Mythmaking*, Baltimore and London: The Johns Hopkins University Press.
Usscher, R. G. (1973) *Aristophanes' Ecclesiazusae*, Oxford: Oxford University Press.
Vellacott, P. (1975) *Ironic Drama. A Study of Euripides' Method and Meaning*, Cambridge: Cambridge University Press.
Vernant, J.-P. (1981a) 'The myth of Prometheus in Hesiod', in R. L. Gordon (ed.) (1981). (First published as 'Le Mythe prométhéen chez Hésiode' in J.-P. Vernant (1974) *Mythe et Société en Grèce Ancienne*, Paris: Maspero.)
Vernant, J.-P. (1981b) 'Sacrificial and alimentary codes in Hesiod's myth of Prometheus', in R. L. Gordon (ed.) (1981). (First published as 'Sacrifice et alimentation humaine à propos du Promethée d'Hésiode', *Annali della Scuola Normale di Pisa* 7, 1977.)
Vernant, J.-P. (1982) *Myth and Society in Ancient Greece*, London: Methuen. (First published as *Mythe et Société en Grèce Ancienne*, Paris, Editions Maspero, 1974.)
Vidal-Naquet, P. (1981a) 'Slavery and the rule of women in tradition, myth and utopia', in R. L. Gordon (ed.) (1981). (First published as 'Esclavage et gynécocratie dans la tradition, le mythe et l'utopie', *Recherches sur les structures sociales dans l'Antiquité classique*, Paris: CNRS.)
Vidal-Naquet, P. (1981b) 'The Black Hunter and the origin of the Athenian *ephebeia*, in R. L. Gordon (ed.) (1981). (First published as 'Le Chasseur noir et l'origine de l'ephébie athénienne', *Annales ESC* 23 (1968).)
Vogt, J. (1974) *Ancient Slavery and the Ideal of Man*, Oxford: Basil Blackwell.
Wolff, H. J. (1944) 'Marriage law and family organization in ancient Athens', *Traditio* 2: 43–95.
Wright, F. A. (1923) *Feminism in Greek Literature*. Reprinted 1969, Port Washington, New York: Kennikat.
Wyse, W. (1904) *The Speeches of Isaeus, with Critical and Explanatory Notes*, Cambridge: Cambridge University Press.

INDEXES

PASSAGES QUOTED FROM GREEK SOURCES

Numbers in italic refer to the original sources, those in roman to the pages on which they are quoted in this book.

GENERAL INDEX

The index refers to the text only.